IEE CIRCUITS, DEVICES AND SYSTEMS SERIES 11

Series Editors: Dr D. G. Haigh
Dr R. S. Soin
Dr J. Wood

Selected topics in advanced solid state and fibre optic sensors

Other volumes in the Circuits, Devices and Systems series:

Selected topics in advanced solid state and fibre optic sensors

Edited by
S. M. Vaezi-Nejad

The Institution of Electrical Engineers

Published by: The Institution of Electrical Engineers, London,
United Kingdom

British Library Cataloguing in Publication Data

A CIP catalogue record for this book
is available from the British Library

ISBN 0 85296 779 9

Printed in England by TJ International Ltd., Padstow, Cornwall

Contents

Preface

Whilst the motivation for producing this book has been to provide a textbook for postgraduate students, the breadth of coverage makes it suitable as a supporting textbook for undergraduate courses in sensors, measurement and instrumentation. The book is also of relevance to practising engineers who wish to update their knowledge of fibre optic and solid state sensors. Fibre optic sensors use optical fibre either as the sensing element or simply as a waveguide. These sensors exploit the benefits of optical fibres: immunity to electromagnetic interference, high sensitivity, flexibility, good electrical isolation, easy multiplexing, and the ability to operate in different environments. Solid state sensors are defined here as sensors based on solid state materials such as crystalline semiconductors, polymers, amorphous materials and thin films. In these sensors, solid state materials are responsible for the sensor operation. For example, in integrated sensors crystalline semiconductors are chiefly responsible for the sensing action. However, this type of sensor is different from other solid state sensors by virtue of its small size and by the techniques used in its manufacture.

The word 'advanced' is used in the title to emphasise not only the sensors which have been discovered in recent years but also that significant research and development activities are still in progress. The authors of each chapter are experienced academics who are currently carrying out research and development work on the particular type of sensor treated in their chapter. As well as describing the principles of operation, characteristics and application of the sensors under consideration, recent advances and future prospects are also discussed.

Chapter 1 deals with the relevant background and sets the scene for the following six chapters. Chapter 2 describes instrumentation systems based on optical techniques, which are often found as associated or integral parts of sensing systems. It also provides basic background for the engineer who is not necessarily familiar with the processes of interactions of light with matter. Fibre optic sensors and recent developments in amplitude, wavelength, phase and polarisation modulating sensors are described in Chapter 3. Advances in amorphous semiconductor photoreceptors and x-ray microbiological applications, electrically conducting polymers for sensing volatile chemicals and thin film (AlClPc) phthalocyanine gas sensors are the subjects of Chapters 5, 6 and 7 respectively.

I am grateful to all of the authors for their contributions; many of them were also pre-occupied with research, teaching and supervision responsibilities at the time of writing. I would like to express my sincere thanks to my family, my postgraduate students and colleagues who helped to bring this book about. It would be wrong to conclude this preface without special mention of Mr. Jonathon Simpson, former commissioning editor of the IEE, and Dr. Robin Mellors-Bourne, Director of Publishing and Ms Sarah Daniels, Book Production Editor of the IEE, UK, for their support and useful comments.

S. M. Vaezi-Nejad

Contributors

Dr D W E Allsopp
Lecturer
Department of Electronic and Electrical Engineering
University of Bath
Bath BA2 7AY, UK

Dr M E Azim-Araghi
Former member of Advanced Materials & Photonics Group
School of Physics and Chemistry
Lancaster University
Lancaster LA1 4YB, UK

Dr W B Betts
Lecturer
Department of Biology
University of York
Heslington
York YO10 5DD, UK

Dr S Hadjiloucas
Lecturer
Department of Cybernetics
University of Reading
PO Box 225, Whiteknights
Reading RG6 6AY, UK

Dr D A Keating
Lecturer
Department of Cybernetics
University of Reading
PO Box 225, Whiteknights
Reading RG6 6AY, UK

Dr A Krier
Senior Lecturer and Head of Advanced Materials & Photonics Group
School of Physics and Chemistry
Lancaster University
Lancaster LA1 4YB, UK

Dr K C Persaud
Senior Lecturer
Department of Instrumentation & Analytical Sciences
University of Manchester Institute of Science & Technology
PO Box 88, Sackville Street
Manchester M60 1QD, UK

Dr Siswoyo
Researcher
Department of Instrumentation & Analytical Sciences
University of Manchester Institute of Science & Technology
PO Box 88, Sackville Street
Manchester M60 1QD, UK

Dr S M Vaezi-Nejad
Reader
Electronics Research Group
Advanced Instrumentation & Control Research Centre
Medway School of Engineering
University of Greenwich
Chatham, Kent ME4 4TB, UK

Chapter 1

Introduction

S. M. Vaezi-Nejad

In this chapter after introducing the basics of measurements in Section 1.1, the characterisation and classifications of sensors are described in Section 1.2. Principles of operation and applications of four groups of sensors are included. The first group is common sensors, the majority of which are the well-established solid state sensors routinely used in a wide range of industrial applications. For clarity of presentation, these sensors are identified in accordance with their input measurand as chemical, mechanical, magnetic, biological and optical sensors. The next three groups are identified in relation to their industrial applications as environmental, healthcare and automotive sensors. Classification in a more precise manner would necessitate undesirable inclusion of a huge number of other sensors. The purpose of Section 1.3 is twofold: first, to review selected classes of advanced sensors that offer significant scope for commercial exploitation, and secondly to set the scene for the next six chapters, which focus on industrially important areas of fibre optic and solid state sensors. Advanced sensors, challenges and opportunities are described in Sections 1.4 and 1.5, respectively. This chapter concludes with a review of the remaining six chapters.

1.1 Background

1.1.1 The essence of measurements

There are several terms employed in the measurement field, the most common of which are instrument, accuracy, precision, sensitivity, resolution and error. Definition of such terms is given in Table 1.1. Note that given a fixed value of a variable, precision is a measure of the degree to which successive measurements differ from one another. Errors may come from different sources and are usually classified as gross errors,

Table 1.1   Common terms used in measurement work

Term	Definition
Instrument	A device for determining the value or magnitude of a quantity or variable.
Accuracy	The closeness with which an instrument reading approaches the true value of the variable being measured.
Precision	A measure of the reproducibility of the measurements.
Sensitivity	The ratio of output signal or response of the instrument to a change in measured value.
Resolution	The smallest change in measured value to which the instrument will respond.
Error	Deviation from the true value of the measured variable.

systematic errors and random errors [1]. Gross errors are largely human errors, among them misreading of instruments, incorrect adjustment and improper application of instruments, and computational mistakes. Systematic errors are shortcomings of the instruments such as defective or worn parts, and effects of the environment on the equipment or the user. Random errors are due to causes that cannot be directly established because of random variations in the parameters of the system of measurement.

The metric system with which the reader should be familiar provides a modern set of measurement units. For example, the fundamental units of length, mass, time and temperature are the metre (m), kilogram (kg), second (s) and Kelvin (K), respectively. Units derived from the fundamental units are area (m^2), volume (m^3), velocity (m/s), force (kg m/s), etc. A standard of measurement is a physical representation of a unit of measurement. Just as there are fundamental and derived units of measurement, we find different types of standards of measurements classified by their functions and applications in four categories: international standards, primary standards, secondary standards and working standards [2].

International standards are defined by international agreement and represent certain units of measurement to the closest possibly accuracy that production and measurement technology allow. Primary standards are maintained by national standards laboratories in different parts of the world, but secondary standards are the basic reference standards used in industrial measurement laboratories. Working standards are the basic reference standards used in industrial measurement laboratories. A statistical analysis of measurement data is common practice because it allows an analytical determination of the uncertainty of the final test result. Statistical analysis may involve the arithmetic mean, deviation from the mean, average deviation and standard deviation, using the well-known mathematical expressions [3].

1.1.2 Measurement systems

Measurement systems are of great importance in a wide variety of domestic, industrial, medical and military activities. There has been a significant growth in the number and sophistication of instruments used in measurement systems. From the application point of view, measuring instruments can be divided into three major areas of regulating trade, monitoring functions and automatic control systems [4]. Instruments used in the regulation of trade measure physical quantities such as length, volume and mass in terms of standard units. Measuring instruments for monitoring purposes provide information which enables us to take some prescribed action [5].

In simple cases, a measuring instrument consists of a single unit, which gives an output signal according to the magnitude of the unknown variable applied to it. More complex measurement instruments consist of several separate elements such as sensors, converters, signal processing, signal transmission, etc. Table 1.2 shows typical elements of a measurement system [6].

Table 1.2   Typical elements of a measurement system

Elements of a measurement system	Examples
Sensing elements	Sensors.
Signal conditioning elements	Detection bridges, amplifiers, ac carrier systems such as current transmitters, oscillators and resonators.
Signal processing elements and software	A/D converters, microcomputer systems, microcomputer software, signal processing calculations.
Data presentation elements	Pointer scale indicators, analogue chart recorders, small scale alphanumeric displays, large scale displays, digital printers.

1.2 Characterisation and classification of sensors

Sensors are essential elements of any measuring system. The word sensor comes from the Latin 'sentire' meaning 'to perceive'. In practical terms this means a device that responds to a physical or chemical stimulus such as heat, light, pressure, sound, magnetism or motion. Sensors are therefore energy transformers capable of converting a stimulus into an electrical output quantity that may have to be processed in accordance with a given algorithm. Another word closely related to sensor is 'transducer'. Sensors and transducers are sometimes used as synonymous terms. The differences between them are very slight. A sensor performs a *transducing* action, i.e. converting energy from one form to another, and the transducer must necessarily sense some physical or chemical signals. In this book the word 'sensor' is used for a device that detects or measures an input signal, and the word transducer refers to a device which performs subsequent transduction operations in a measurement or control system. For example, consider the general sensing system illustrated in Figure 1.1.

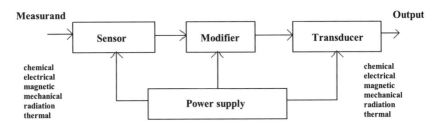

Figure 1.1   A general measurement system

The measurement is detected by the sensor, the output of which is then applied to a modifier. Normally the sensor output is an electrical signal that is processed by the modifier before it is fed into a transducer. The modifier may also simplify, attenuate, modulate or demodulate the signal. The function of the transducer is to connect the signal into a form suitable for display, recording or performing some action. Technological aspects used in evaluating sensor characteristics and to compare sensor performance are listed in Table 1.3 [7].

It is not an easy task to classify sensors in a very specific manner as there are so many types, each operating on different physical principles. As shown in the Appendix (Section 1.7), in the UK alone, the variety of sensors and their wide range of applications have resulted in the formation of several sensors and sensor related interest groups, clubs and associations. The sensors treated in the next six chapters are classified as fibre optic sensors and solid state sensors.

1.2.1 Common sensors and their applications

As indicated in Table 1.4, there is a wide range of common sensors based on solid state materials [7–48].

Table 1.3   Sensor characterisation

Characterisation aspect	Definition and comments
Ambient conditions	The effect of ambient conditions such as temperature, acceleration, vibration, shock, ambient pressure, moisture, electromagnetic field, etc.
Full scale output (FSO)	The upper limit of sensor output over the measurand range is called the full scale (FS). The algebraic difference between the end points of the output is 50%.
Hysteresis	This is the maximum difference in output, at any measurand range when the value is approached first with an increasing and then a decreasing measurand. It is expressed in percent of FSO during one calibration cycle.
Linearity	The closeness between the calibration curve and a specified straight line.
Measurand range	The value of the measurand over which the sensor is intended to measure, specified by upper and lower limits.
Offset	The output of a sensor, at room temperature, unless otherwise specified, with zero measurand applied.
Operating life	The minimum length of time over which the sensor will operate, either continuously or over a number of on–off cycles whose duration is specified, without changing performance characteristics beyond specified tolerances.
Output format	This is a function of the measurand that can be in analogue or digital form.
Overload characteristics	This is the maximum magnitude of the measurand that can be applied to a sensor without causing a change in performance beyond the specified tolerance.
Repeatability	The ability of a sensor to reproduce output readings at room temperature, unless otherwise specified, when the same measurand is applied to it consequently, under the same conditions and in the same directions.
Resolution	The minimal change of the measurand value necessary to produce a detectable change at the output.
Selectivity	The ability of a sensor to measure one measurand in the presence of others.
Sensitivity	The ratio of the change in sensor output to the change in the value of the measurand.
Detectivity	The least input measurand that can be detected by a sensor.

Table 1.3 – continued

Characterisation aspect	Definition and comments
Speed of response	The time at which the output reaches 63% of its final value in response to a step change in the measurand.
Stability	The ability of a sensor to maintain its performance characteristics for a certain period of time.

Table 1.4   Some common examples of sensors and their applications

Type of sensor	Typical applications	References
Chemical	Fuel composition sensors, field effect chemical sensors, thick-film electrochemical sensors, clinical analysis, environmental and process control.	7–14
Acoustic	Ultrasonic sensors, environmental gas sensing, surface acoustic wave microsensors, corrosion sensors.	5, 10, 15
Mechanical	Pressure sensors, strain sensors, silicon accelerometer, capacitive sensors, piezoresistive sensors, mechanical shock measurement.	16–23
Magnetic	Hall sensors, magneto-transistors, thermogalvanomagnetic sensors, CMOS magnetic field sensors.	24–29
Biological and biomedical	Electrochemical sensors, ion-sensitive field effect transducers (ISFET), electrocardiography, integrated biosensors for on-line monitoring and control.	30–38
Optical	Photodiodes, CCDs, phototransistors and colour sensors.	39–47
Thermal	Thermopile, infra-red detectors, thermal and temperature sensors.	48

Chemical sensors cover a wide range of applications including clinical analysis, environmental monitoring and process control. For example, an enzyme electrode is a device in which the activity of an enzyme, or group of enzymes, is monitored electrochemically. The abundance of pesticides in the aquatic environment has become a matter of major concern as large quantities of herbicides which are being applied to crops world-wide eventually find their way into rivers, lakes and seas, causing damage to many forms of aquatic life. Amperometric enzyme electrodes may be used for constant monitoring of the herbicide atrazine. There is increasing demand for humidity-sensing elements to be used for automatic humidity control in a variety of technological fields such as the electronics industry, drying machines, food manufacturing processes and combustion controlling systems. The humidity sensitive materials may be electrolytes, organic polymers or porous ceramics. Various ceramic materials have been studied as humidity sensors due to their mechanical strength, temperature resistance and chemical stability. Particular attention has been focused on clarification of the relationships between humidity sensitivity, surface absorption phenomena and the micro-structure of the porous ceramic sensor element.

Mechanical sensors include pressure sensors, silicon accelerometers and ion sensitive field effect transistors (ISFET). Pressure sensors using flexible diaphragms of silicon have been available for some time. Pressure causes a static deflection of the diaphragm, which is measured by piezo or capacitance methods. There is fundamental interest in this type of sensor because of its inherent accuracy and digitally compatible output. The ISFET for the measurement of pH is very promising. The most common membranes for pH-ISFET gates are organic materials such as silicon dioxide (SiO_2), silicon nitride (Si_3N_4), alumina (Al_2O_3) and tantalum oxide (Ta_2O_5). Si_3N_4 seems to be a very good choice due to its sensitivity, long term stability and high resistance to hydration. This material is also well known in microelectronics, and is extensively used in complementary metal-oxide-semiconductor (CMOS) technology. Zirconia sensors of the fully sealed miniature pump-gauge type have applications including monitoring and control of the air-to-fuel ratio in combustion systems, life support systems at standard or varying barometric pressures and general environmental management.

Magnetic sensors are normally used in the measurement of the magnitude and direction of the magnetic field using the laws and effects of magnetic or electromagnetic fields. Technological advances in the area of thin-film microelectronics have allowed magneto-resistive sensors to be developed that are suited for the mass market in view of their low cost high sensitivity and low power consumption aspects. Both bipolar and MOS technologies have been used for fabrication of magneto-transistors. In these devices, the magnetic field modifies the carrier-transport conditions and produces changes in their electrical output characteristics that are normally negligible. The design and operating conditions of the device can be optimised with respect to the magnetic field sensitivity of its characteristics. Magneto-resistors are sub-divided into two categories, namely geometric and physical devices. In geometric magneto-resistors, geometrical factors such as the length, width, thickness and shape of the material cause the resistance change. Physical magneto-resistors involve such processes as changes in the velocity distribution of free charge carcarriers in the material. Optical fibre magnetic field sensors are generally based on either light intensity modulation or light phase modulation.

Figure 1.2 shows the components of a typical intensity modulation sensor. These sensors are generally associated with displacement or some other physical perturbation that interacts with the fibre or a mechanical transducer attached to the fibre. For example, the transmissive fibre optic sensor is used to measure the magnetic field and to determine the current if a magneto-optical material is placed in the light path. Phase modulation sensors compare the phase of light in a sensing fibre to that of a reference fibre in an interferometer. A silicon fibre is embedded in a magneto-resistive jacket that expands or contracts in the presence of a magnetic field. Ideally, the magneto-resistive coating is in fibre form, but bulk materials can be used such as stripes or cylinders with the fibre bonded to the material. Normally, the frequency response of a phase-modulated magneto-strictive fibre optic sensor is dependent on the material as well as the geometry, but above approximately 1 kHz, a general decrease in sensitivity is observed. The operating principle of magneto-diodes is based on the three phenomena of carrier injection, the Hall effect and surface recombination or generation of carriers.

Typical relative sensitivities of these sensors are a few volts per mA at a bias current of a few mA. Magneto-transistors are bipolar transistors whose structures and operating conditions are optimised with respect to the magnetic-field sensitivity of the collector current. Other examples of magnetic sensors include CMOS magnetic field sensors and thermo-galvanomagnetic sensors.

Biological sensors or biosensors are basically biological receptors such as enzymes, micro-organisms or antibodies, which are coupled to an electronic transducer for converting the biological input signal into an electrical signal. These biological receptors can either be immobilised on the transducer surface, or on a separate membrane or on carrier particles, in which case the receptor signal reaches the transducer by sample flow.

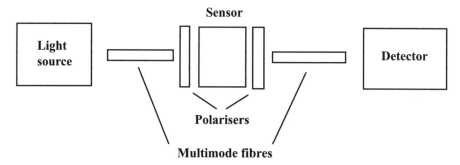

Figure 1.2   The basic components of a typical intensity modulation sensor

Most biosensors are based on electrochemical devices such as the Clark cell for their detection or transduction element. However, recent developments have focused on chemically sensitive semiconductor devices, fibre optics, thermistors, chemically treated electrodes and many other microelectronic devices. One of the most important applications of biotechnology for the 21st century is biodegradable polymer-based scaffolds for the regeneration of human nerve, liver and bone tissue. Carbon dioxide processing of biodegradable polymer seems very promising for the production of scaffolds with biominetic surfaces and growth hormones incorporated into their structure [49]. The process is environmentally friendly and may also cut the cost of scaffold production.

Table 1.5 shows some of the more common types of biosensing process used [31, 34]. Significant laboratory research is focused on glucose sensors because of their application in the detection/control of glucose concentrations in diabetic patients. There are simple and practical methods for fabricating glucose sensors under laboratory conditions as used in many research centres. The performance of such sensors is still under investigation but they have been used successfully on rabbits in glucose control tests. Biosensors based on potentiometric detection are constructed as a field effect transistor with the metal gate replaced by an ion sensitive membrane, a solution and a reference electrode. There are problems of drift associated with the device leakage currents that have to be resolved. There are several types of biosensors for on-line monitoring biotechnological processes. Lack of instrumentation with the required specificity and sensitivity for process monitoring and control seem to be particular problems for production processes. Better hardware and software, including closed loop expert systems for real time fault detection and diagnosis, are highly desirable.

Table 1.5   Common types of biosensing processes

Analyte	Biological sensing element	Transducer type
Enzyme substrates (e.g. glucose, urea)	Tissues Cells	Optical
Metabolic chemicals (e.g. oxygen, ethanol)	Organisms	Electrochemical: (a) potentiometric (b) amperometric (c) impedometric
Antigens/antibodies Nucleic acids (e.g. DNA)	Receptors Antibodies	Acoustic (mass)
Ligands (e.g. hormones) (thermal)	Enzymes	Calorimetry

Optical sensors are devices in which an optical signal is changed in some reproducible way by an external stimulus such as temperature or strain. Optical sensors include several devices, because an optical beam is characterised by several variables such as intensity, wavelength spectrum, phase and state of polarisation. Visible light colour sensor charge coupled devices (CCDs), photodiodes, phototransistors and image intensifier tubes have advanced considerably in recent years [50, 51].

Optical sensors, which use optical fibre, are called fibre optic sensors. These sensors are classified as extrinsic or intrinsic according to their mode of operation. In extrinsic fibre sensors, the fibre itself is passive, and is used only to transmit light from one location to another. The fibre plays no role in the sensing; sensing results from non-fibre related effects. Any influence of the fibre itself on the sensing is undesirable and will lead to inaccurate sensor response. In intrinsic fibre sensors, the outside environment can affect light in the optical fibre by modulating the light intensity, frequency, polarisation, phase or other parameters of the fibre. As it will be seen in Chapter 3, each of these effects can be the basis for a family of fibre sensors.

The use of fibre optic technology with smart structures enables the practical realisation of multiplexed arrays of fibre sensors interconnected by other optical fibres. Multiplexed arrays of sensors allow vast quantities of sensory data to be gathered from all over the structure, and will therefore require complex signal-processing techniques, for example employing neural networks, to process the data. Fibre optic technology also enables the implementation of distributed sensors which are able to measure the value of a parameter continuously along the length of a single optical fibre. Fibres as long as several kilometres can be used as distributed sensors.

Integrated optics involves the concentration of light in thin film waveguides that are deposited on the surface or inside a substrate. Dielectric light waveguides can be extremely small in their dimensions because of the short wavelength of the light. The reduced size of integrated optical circuits makes it possible to achieve a very high density of components. An additional advantage of the small size and rugged construction of dielectric light waveguides is their insensitivity to vibrations and temperature changes in their environment. Integrated optical circuits can be made two dimensional if the light is allowed to spread out and propagate in the so called slab waveguide, which is basically a thin dielectric layer on top of the glass substrate. These circuits become one dimensional if the dielectric layers are deposited as thin strips that guide not only in the dimension of the substrate surface but also confine it to a narrow region defined by the stripe waveguide [52, 53]. Although signal transmission using light waves is now well-established, the optical signal usually has to be converted back into an electrical one if any processing needs to be done. Integrated optics technology should incorporate as much signal processing as possible on the optical signal itself [54, 55].

Research and development work in the field of integrated optics includes application of a family of optical and electro-optical elements in thin film planar form for assembly of a large number of such devices on a single substrate. Such systems would have similar advantages to those gained by the adoption of the basic idea of the integrated circuit in microelectronics [56]. Technological developments involving point and distributed sensors using optical fibre or integrated optics are summarised in Table 1.6. These sensors basically consist of a source of light such as a light emitting diode (LED) or a laser diode connected to the fibre or integrated optic chip as the waveguide. As mentioned earlier, this category of sensors offers considerable advantages compared to electrical

Table 1.6   Fibre optic and integrated optic sensors technologies [57–61]

Physical parameter or type	Examples and principles	Comments
Interferometers	Integrated optic displacement sensor.	Medium investment. Medium production and cost.
Wavelength modulation	pH sensors in which chemical reactions are measured by calorimetric changes.	Very low level of miniaturisation. High flexibility.
Intensity modulation	Force sensor based on microbending losses.	Hybrid integrated optics circuits.
Polarisation modulation	Electric current sensors based on Faraday effect.	Discrete fibre optics components.
*Chemical	pH and phosphorus dioxide (PO_2) measurements; pollution monitoring.	Expensive instrumentation (above comments apply to the integrated optic displacement sensor and sensors based on modulation of wavelength, intensity or polarisation).
*Distributed sensors	Optical time domain reflectometer.	
*Resonant	Optically excited resonant sensors employing a micromachined structure set in resonance by a pulse light beam.	
*Non-contact displacement	For metrology, industrial inspection and monitoring applications.	

* Most promising areas of research activity with good market opportunities.

sensors, but such advantages are offset against the increased cost of the instrument. It should be noted that the most promising areas of research activity are also identified in Table 1.6.

1.2.2 Environmental sensors and recent advances

The impact of pollution on the environment has become a major public and political concern. Global environmental monitoring and pollution control issues include acid rain, ozone depletion and the greenhouse effect. These and more localised issues, such as nuclear waste, sewage contamination of beaches and dissolved metals in water courses, have resulted in growing public pressure for the introduction of new specific legislation. There is a diversity of environmentally hazardous chemical, physical and biological quantities produced by numerous processes and industries. According to a study commissioned by the UK Department of Trade and Industry (DTI), there are a number of important pollutants such as carbon dioxide (CO_2), radon or methane and hydrogen sulphide (H_2S) from approximately sixty major sources including paper making, polymer processing, diesel engines, car exhaust gases and particulates [62].

Techniques to minimise environmentally hazardous emissions and phenomena, methods to control and monitor the effectiveness of such techniques, and monitoring the environment itself will be required for the foreseeable future. Reliable advanced sensors offering high sensitivity and selectivity for ambient air quality monitoring which is principally concerned with particulates, carbon monoxide (CO), sulphur dioxide (SO_2), ozone (O_3) and nitrogen dioxide (NO_2) are highly desirable. As shown in Table 1.7, a number of sensors from the two broad categories of solid state sensors and guided wave optical sensors have been under investigation in the UK for industrial exploitation [63].

Solid state materials such as sintered masses of polycrystalline tin (IV) oxide SnO_2 are widely used for the detection of flammable or toxic gases. Semiconductor gas sensors have become a well-established component of the technology available for monitoring atmospheric composition.

A variety of reactions have been described as mechanisms by which the resistance of such sensors can be modulated, for example, reaction with surface oxygen species, responses to changes in moisture content of the atmosphere, surface ion exchange and reactions involving precious metals. The resistance of semiconductor oxides may depend on atmospheric composition via one or more of several mechanisms. Thus in order to achieve optimal performance of semiconductor gas sensors in terms of stability and selectivity, it is necessary to exercise control over these mechanisms separately. Solid state glass electrodes fabricated with a transition layer between an ion sensitive glass and a metal contact show a linear response to pH and improved performance regarding range

Table 1.7   Advanced sensors for environmental monitoring and control

Category	Topics under investigation
Solid state sensors	– Micromachined oscillating silicon structures – Silicon biosensors – FET sensors – Polysilicon sensors – Thin film semiconducting organometallic phthalocyanines (e.g. cupc, pbpc, etc.) and polypyrole – Langmuir–Blodgett (mono-molecular) thin films – Gas sensitive organic films such as: nitrogen oxide gases (NO_x), ammonia (NH_3), chloride (Cl_2), hydrogen chloride (HCl) and inorganic metal oxide films (for O_3, toxic species, etc.) – Thick film sensors for mechanical (e.g. load, strain, pressure) and chemical (e.g. metallic ions, toxic, gases, etc.) quantities
Guided wave optical sensors	– Monomode (interferometric) sensors – Multimode sensors – Distributed sensors – Fibre optic instruments and analysis – Sensor multiplexing techniques – Novel sensory fibres and other optical components – Hybrid fibre optic/silicon (or SiO_2) resonator sensors – Integrated optical sensors

and drift of standard potentials compared to those fabricated with metal-only contacts. Although the fast response of heat-treated phthalocyanine films is a significant advantage in the development of simple NO_2 sensing instruments, it has been found that some slow response components remain, and that the operating cycle of the sensor must be carefully designed to avoid the gradual dominance of these components. Such sensors are discussed in detail in Chapter 7. It has been demonstrated that an amperometric zirconia oxygen gas sensor can be fabricated using screen-printing techniques. However, careful attention to ink formulation and sintering conditions is essential to achieve a low porosity electrolyte film with no loss of adhesion from the substrate or cracking of the film. One of the active areas in gas sensor research is in the use of sensor arrays based on semiconductors or mass sensitive devices coupled with pattern recognition techniques to provide some degree of selectivity.

Recent progress made in environmental monitoring and pollution control includes introduction of ozone monitors in the market, capable of accurately measuring ozone concentration across a wide range from 0.001 ppm to 99,000 ppm. Solid state Doppler radars (1–2 μm) and CO_2 (9–11 μm) coherent laser systems have been used for a wide range of atmospheric remote sensing applications. A significant amount of work is being conducted in radar systems towards improved engineering and reliability. The capability of these systems is especially suited to detecting and measuring wind shear for ground-based airport surveillance and commercial airline applications. The measurement of the global wind field has enormous potential for contributing to global weather forecasting accuracies. The detection and tracking of aircraft wake vortices at airports in order to increase airport capacity as well as safety is another useful application [64]. Laser radar, also known as light detection and ranging (LIDAR) or laser detection and ranging (LADAR), is utilised for remote measurement of the properties of hard and soft (atmospheric) targets. Laser radar systems combine the principles of microwave radar and optics. High frequency operation (30–1000 Hz), coupled with excellent spatial and temporal coherence, results in the laser providing diffraction limited operation, large aperture gain, high intensity, illumination at long range, a photographic size spatial footprint at the target, and coherent detection receiver operation [65].

As the number of vehicles on the road and the consequent demand for transportation fuels continues to rise at a steady rate throughout the world, there is a growing concern with pollution and greenhouse gases. Increasingly stringent restrictions are being placed on the amount of polluting gas that a car exhaust can emit. One of the most established methods of inspecting combustion conditions in the car cylinder is based on pressure measurement. Low cost and reliable pressure sensors such as those based on thick film technology are very desirable. Low cost planar amperometric zirconia oxygen sensors have been fabricated by screen-printing electrodes and electrolyte on an aluminium substrate. Such sensors cost substantially less than potentiometric sensors as they are constructed from discrete components using planar technologies. High temperature gas sensors that can be operated over a large temperature range might also be a robust new sensor technology in the 'electronic nose', with a versatile response pattern to a large number of different molecules. Silicon carbide based high temperature gas sensors are a possibility here. Since silicon carbide allows a good quality oxide to be grown on it, it is possible to process MOS devices from this material [66, 67].

Deregulation and tougher clean air rules in the USA are forcing electricity generating companies to modernise their technology. A recent publication by the Instrument Society of America has identified a number of critical sensing needs, including:

- Measuring pulverised coal flow to individual burners to enable burner balancing for improved plant efficiencies and reduced emissions.
- Measurement of NO_x production at the individual burners to enable better emission performance strategies.
- Non-invasive technologies, such as optics for measuring strain in a robust reliable manner.
- Essentially drift-free sensors that do not require frequent, costly calibrations to maintain the required accuracy.
- Early detection of insulation system or mechanical operation failures of power transformers and circuit breakers.
- Sensors for measuring the remaining life of power transformers and circuit breakers, as well as to detect faults and combustible or dangerous gases in vaults and other underground constructions.
- Sensors for measuring the remaining life of wood, metal overhead transmission structures and underground cables in high voltage transmission.
- Voltage and current sensors for monitoring a variety of system conditions, offering advantages of accuracy and low installation and maintenance costs.

Water monitoring and treatment

The importance of water to industry should not be underestimated. In recent years testing kits for boiler and cooling waters have been produced to make industrial water monitoring easy and cost effective. A new water treatment process has been developed in the UK that combines ultraviolet light treatment with chemical oxidation to break down organic pollutants in water, which are difficult to remove using conventional methods. However, the use of magnets in water treatment is an attractive alternative to chemical treatment. Dissolved metals in water courses, leaches from land fill sites, numerous chemical substances, physical phenomena and hazardous micro-organisms show the importance of the need for (i) minimisation of the use of toxic substances and environmentally hazardous practices, (ii) control of the release of pollutants and (iii) monitoring the presence of pollutants in the environment.

There are two principal methods for producing fresh water from brackish or seawater: evaporative types using thermal energy and membrane types using electrical energy. Thermo-compression distillation and reverse osmosis desalination are common desalination process techniques both requiring chemical treatment for increase in potability of the water. Thermo-compression distillation produces water that is too pure for use as potable water, and blending with raw water or chemical dosing is necessary to make it palatable. With reverse osmosis, chemical dosing is important to ensure that the water quality complies with World Health Organisation recommendations. As the above treatments are costly and environmentally intrusive, more cost effective and environmentally less intrusive methods of conserving water are desirable [65, 68, 69]. As far as the market opportunity in the water industry is concerned, there is a need for a reliable method of predicting if a particular water supply is suitable for treatment to prevent scaling by using a method other than chemicals or a traditional water softener.

1.2.3 Sensors for healthcare applications

The rare combination of public demand, a technology-oriented end-user and rapidly developing technology have resulted in a growing range of sensor and instrumentation products for health care based on new technologies [70]. Table 1.8 shows how market opportunities can be divided according to the site of use of the product into centralised and decentralised sites. There are three basic enabling technologies appropriate to the healthcare field: biotechnology, guided wave optics and microelectronics.

1.2.4 Automotive sensors and their applications

Automotive sensors currently in use are shown in Table 1.9. Electronic sensors have been used for measuring the crankshaft position for ignition timing purposes. Electrical methods such as potentiometers are used for many automotive applications including fuel level measurement, sensing throttle and brake pedal positions, steering wheel motion and suspension displacement and automatic gearbox control. Potentiometer sensors are cheap and reasonably reliable but they suffer from wear as they rely on a sliding contact. Measurement of speed of the vehicle, its oil pressure, airflow into its engine and ignition timing are also important. Modern timing sensors are based on electromagnetic, Hall effect or optical methods. Once the engine load is derived from a pressure sensor and its speed is measured by a speedometer, the required ignition timing can be determined.

From the measurement of the amount of oxygen in the exhaust, the air–fuel ratio entering the engine is controlled to ensure complete combustion. A three-way catalytic converter is placed in the exhaust to remove the critical pollutants of CO, unburnt hydrocarbons (HC), and NO_2. Vehicle temperature measurement is essential for several control functions, particularly that of electronic engine management, where the inlet air temperature is required. For applications such as tyre condition, and propshaft joint monitoring, non-invasive temperature measurement techniques and infra-red emission sensors are most suitable.

One of the most widely used types of sensors in automotive engineering is the displacement sensor. Potentiometers are frequently used for sensing brake pedal position, steering wheel motion and suspension displacement. Ultrasonic displacement transducers may be used for short-range collision warning and fuel tank level sensing. Gas and fuel

Table 1.8 Market classifications and opportunities in healthcare sensors [49,70]

Classification	Identified site	Comment/typical sensors	Advanced sensor and opportunities
Centralised sites	Hospital bedsides	Temperature, blood pressure, heart rate, blood gases	Fibre optic based blood pressure transducers
	Emergency room	Blood glucose level of drugs of abuse, alcohol and in pregnancy	High speed of response and disposable sensors
	Operating theatre	Heart rate monitoring, blood pressure monitoring, respiration and blood oxygen	Oximeters and blood gas monitors
	Centralised diagnostic facilities	Infectious diseases detection, blood analysis, drug monitoring	High speed biosensors
	Therapy rooms	Sensors for monitoring, radiation therapy and physical therapies	Disposable probes
	Medical imaging*	X-ray, CT scanner, ultrasound and NMR facilities	Systems-based fibre optic sensors
Decentralised sites	Consulting rooms	Peak flow meters, electrocardiographs, tonometers, audiometers	Analysers for spot checks
	Mobile emergency	This is not a significantly large market. Example: pulse oximeter	Small size, low weight and portable equipment
	Home, supervised	Sensors for monitoring home dialysis, oxygen therapy drug treatments for diabetes, asthma and epilepsy	Drug therapy monitoring
	Home, unsupervised	Very rapid growth market. Example: self diagnostic products such as pregnancy and ovulation prediction tests, glucose monitoring kits, faecal occult blood tests	Fertility testing. Glucose test kits

* See also Chapter 4 section 4.4.

Table 1.9   Automotive sensors and their uses

Type of sensor	Typical application	References
Gas and fuel composition	Exhaust gas oxygen (EGO) sensors, selective gas sensors, lean burn oxygen sensors	71–73
Airflow	Hot-wire mass airflow meters, corona discharge mass airflow sensors, ultrasonic mass airflow sensors	74, 75
Temperature	p–n junction sensors, thermocouples, bimetallic temperature sensors, resistive temperature transducers, liquid crystal temperature sensors	76, 77
Position, displacement and velocity	Resistive position detectors, potentiometer sensors, eddy current displacement transducers, inductive, capacitive and ultrasonic displacement transducers, variable-coupling transformers, variable reluctance transducers, Hall effect sensors	78–80
Liquid level	Optical liquid level sensors, capacitive liquid level sensors, thermal liquid level sensors	81, 82
Pressure	Elastic pressure sensors, strain gauges and piezoresistive sensors, capacitance pressure sensors, silicon micromachined pressure switches	83–85
Combustion	Ionisation sensors, in-cylinder pressure sensors, optically sensed diaphragms, pressure sensors, knock sensors	86, 87
Shock and vibration	Piezoelectric accelerometers, piezoresistive accelerometers, vibration velocity sensors, capacitive accelerometers, inertia switches	88, 89
Torque	Strain gauge torque transducers, ratio metric torque sensors, brake dynamometers, torsion-bar torquemeters, magnetostrictive torque sensors	90, 91

composition sensors continue to play a major part in improving the performance of the motor vehicle. Lean burn sensors are being recognised as the best way of improving both emissions and fuel consumption. Manifold absolute pressure (MAP) sensors based on capacitive, inductive, potentiometric and strain gauge techniques are commonly used for automotive pressure measurement. Devices based on thick film technology offer a cheap but larger size pressure sensor compared to silicon. Piezoelectric and piezoresistive accelerometers are used to measure shock and vibration. If low frequency measurements are required, the latter is preferred. For high temperature of operation, the centre-mounted compression piezoelectric accelerometer should be used [71, 72]. There are several torque sensors in use. Strain gauge systems using sliding contacts or telemetry are adequate for development work but prohibitively expensive for use on production vehicles. The torsion-bar torquemeters are more suitable as they are cheaper and they avoid the need for telemetry.

1.3 The advanced sensor scene

With the recent explosive growth in sensing and related measurement technologies, it is an impossible task to cover all advances made in sensors [92–100]. However, there follows a general overview of such sensors and the related market opportunities.

A recent report by the Foresight Sensors Action Group in the UK has recognised the growing importance and research opportunities in smart sensors for process industries, biosensors in the food sector, machine vision for manufacturing, thin and thick films as gas sensors, microelec-tro-mechanical (MEM) systems for position sensing or as pressure sensors in the motor industry and fibre optic sensors for temperature measurement and structural monitoring [101]. The technology of MEM systems has grown in complexity and sophistication. Micromachined silicon structures are used as sensors and actuators in the medical, indus-trial, consumer, military, automotive and instrumentation fields. Novel applications are under research involving circuits and functional mechanical structures combined on to a silicon substrate. Such devices would be combined on to smart sensors for integrating directly to microprocessor-controlled equipment [102, 103]. Mobile micro-robots that can be batch fabricated using silicon are an advanced technological application of MEM systems. When developed these could be used in micro-tele-operations for very small inaccessible areas, and in the hand-ling of small biological or electromechanical elements. Shrinking robots and mechanical actuators to the same size as the parts to be manipulated has the advantage that extremely delicate forces can be applied and the relative accuracy of the device can be markedly reduced. Currently, there are few examples of structures that have been manufactured using cost effective integrated circuit batch-fabrication technologies. Most micromachined components are currently more expensive than the microcontrollers to which they are interfaced, but as the new processing techniques are developed for fabrication of both the transducer and the electrical elements, the situation should improve.

The idea of mechanical, civil and aerospace engineering structures which can react or adapt their characteristics in response to signals derived from sensors integrated within such structures has generated considerable interest [104, 105]. For example, strain, temperature, state-of-cure of the bonding epoxies and material damage are the principal measurements to be monitored in composite materials such as those based on carbon and glass fibre. These materials are increasingly used in engineering applications because of their high strength-to-weight ratio. Measurement of strain, vibration, temperature and damage in composite components during manufacture leads to the production of more reliable components. Monitoring of strain or displacement of concrete structures widely used in civil engineering is a matter of some interest [106–108].

Smart or intelligent sensors can interact with the control computer to provide data manipulation. For example, as mentioned in the previous section, optical fibre smart sensors can monitor themselves to assess their structural health using sensing and communication capabilities of fibre optics and electronics. As shown in Table 1.10, there are several such sensors gradually emerging in the market. For application in civil engineering these sensors have been used to monitor crack growth, dynamic loading, length and strain distribution. Systems based on smart sensors, data processing and a man–machine interface enable detection of concealed objects or observation through murky media and checking for shape, appearance or absence of defect. Applications include surveillance, process monitoring, automatic inspection and structural analysis. Noncontacting techniques using X-ray, infra-red and visible light are being investigated for developing such systems. Data processing involves neural networks, knowledge based systems, genetic algorithms, fuzzy logic and other artificial intelligence technologies. Man–machine interfaces include 3D displays, virtual reality, data visualisation and other techniques.

Table 1.10   Optical fibre smart sensors based on fibre microbending effects [109–113]

Principles of operation	Applications
Light absorbing probe, Bragg grating, Rayleigh backscatter, Fabry–Pérot interferometer, Fresnel optical time domain reflectometry (OTDR), Sagnac interferometer.	Civil engineering, earthquake prediction, volcanic action prediction, oil platform movement, telecommunication cable monitoring.

1.4 Challenges and opportunities in solid state sensors

Silicon, thin film and thick film solid state sensor technologies and their applications are summarised in Table 1.11. Note that comparisons of the production, cost, level of miniaturisation, integration and flexibilities of these sensors are also included in the table. Advanced thin film research includes materials improvement and new concepts towards monolithic integration of the electronics. Considerable research effort has been devoted to gas, chemical and electrochemical thick film sensors. Some of the research challenges are [114–124]:

(i)   chemistry and preparation of suitable active layers;
(ii) the understanding of the surface interaction of the active layer with the measured gas;
(iii) ionic conduction phenomena within the active layer;
(iv) preparation of suitable pastes, long term stability and packaging.

Table 1.1.1   *Solid state sensor technologies [126-132]*

Sensor technology	Physical parameter or type	Application/transduction and principles	Comments
Silicon	Mechanical	Pressure sensor/piezoresistive Strain gauge/piezoelectric Accelerometer	• Well-known technology
	Radiant	Solar cell/photoelectric Photodetectors/photoconductive	• Very low cost and high level of miniaturisation
	Magnetic	Position/Hall effect Displacement/magneto-resistive	• Monolithic with low flexibility
	Thermal	Thermocouple/Seebeck effect Thermoresistance/resistivity change	• Modelling and ASIC at early stage
	Chemical	pH-ISFET/ion absorption	• Very high productivity and investment
Thin film	Strain Temperature Pressure	Strain gauges using metal films or semiconductors Thermistors and thermocouples using metals Pressure sensors using flexible diaphragms in metals or glass with deposited strain gauges	• Low cost and medium productivity • High investment • Medium level of miniaturisation
	Humidity	Hygroscopic films such as Ta_2O_5 or polymers measuring electrical conductivity	• Hybrid circuits
	Flow	Thermal flow sensory with bridge network of temperature sensors	• Medium flexibility
	Optical	Photoconductive semiconductor films as optical detectors, image tubes	• Simple production techniques
	Chemical	SnO_2, ZrO_2 films for oxygen, hydrogen and other gas detection	• Film thickness of a few nanometers up to a few microns
Thick film	Temperature Pressure Strain Humidity	Thermal sensors and thermal flow sensors Piezoresistive and capacitive read out	• Film thickness of 10–20 microns • Medium level of production and low cost • Low level of miniaturisation but high flexibility • Hybrid circuits
	Gas	Specialised sensors for avionics	

Solid state sensors, with their ever-improving performance–cost ratios, will be key components with further penetration of microelectronics into new products and new applications [124–132]. The two approaches that can lead to the possibility of integrating semiconductor sensors with microelectronic circuits are sensors in semiconductors and sensors on semiconductors. In the former, semiconductor materials are chiefly responsible for sensor operation, whereas in the latter, alternative materials are deposited on top of the semiconductor substrates to form the sensor. The most important semiconductor for both forms is silicon. The small size of semiconductor sensors not only contributes to their potentially low cost, but also allows them to be integrated with microelectronic circuits. These sensors are realised using a high level of integration, with some or all of the sensor electronics monolithically integrated on the sensor chip itself [123–127].

The so called sensors-on-chip have emerged in the market, offering noticeable advantages. For example, in audio power Darlington transistors with on-chip temperature sensing, through direct connection to the bias circuitry, the temperature sensor can counter the effects of thermal runaway. Being on chip ensures there is no time lag or temperature difference, making for a more stable and reliable amplifier [124]. Sensory voice direct ICs have been implemented as a single chip for low cost speaker dependent voice recognition applications [125]. Other examples include microsensors, thin film multi-sensors, and multi-electrode recording microprobes [126, 127].

High resistivity materials are used in a variety of sensors either as the sensing media or as a thin film and thick film substrate for packaging [128–131]. For example, amorphous semiconductors such as amorphous silicon and its compounds, which have a high absorption in the visible spectral range, are very attractive as colour sensors. Various structures have been suggested, including double Schottky barrier, n–i–p–i–n or p–i–n–i–p layer sequences. Common to all these structures is the problem of achieving a sufficient separation of the signals from the three basic colours that are close to the maximum response of the three types of core cell in the human eye (red, green and blue). Recent work has focused on n–i–p–i–n structures because for this layer sequences of electrons dominate the carrier transport, in contrast to p–i–n–i–p devices where holes with lower drift mobility predominantly determine the electrical performance [132]. Such a structure offers the attractive opportunity of integrating the amorphous silicon based multi-layer with a crystalline ASIC (Application Specific Integrated Circuit) for digital data processing.

It is appropriate to end this section by referring to some of the most important issues that one has to consider very carefully before launching a new sensor into manufacture and commercialisation [117]. These are:

(i) Coping with a new specialised technology which is not widespread. There are unknown factors in manufacture and in use in difficult environments, as well as non-compatibility. Limited suppliers and second sourcing can be a problem for the larger users. It is therefore desirable for the potential sensor manufacturer to work closely with an R&D laboratory or organisation.

(ii) Facing an expensive start-up due to high R&D costs, and expensive production equipment. New sensors have lead times of more than five years, and payback on such investments requires high volume markets. The solution is to share the R&D costs, and small companies should share production facilities.

(iii) Risks associated with unknown factors and performance. Reliability is not proved; temperature and environment ranges might be limited. Assembly and packaging techniques are undeveloped or resolved.

(iv) Non-compatibility problems as the new sensor types may not be standardised. Despite the use of integration of electronics on a silicon sensor chip, some sensors do not provide a microprocessor compatible signal and thus require additional signal processing.

1.5 About this book

It is clear that fibre optic and solid state sensors play a major role in several areas of industry, including transport, health care, environmental monitoring, pollution control, mechanical, civil and aerospace engineering structures and process industry. The following six chapters cover in depth: amplitude, wavelength, phase and polarisation modulating sensors, amorphous semiconductor image sensors, dielectrophoretic sensors, electrically conducting polymers for sensing volatile chemicals and thin film phthalocyanine gas sensors.

Chapter 2 deals with optical sources, detection schemes and recent advances in instrumentation based on the interaction of light with matter. The chapter ends with an overview of recent advances in terahertz (THz) measurement and instrumentation systems based on ultra fast pulses in the far-infra-red region of the electromagnetic spectrum. THz measurement techniques can be used for observation of excited free carrier phenomena in the Si substrate as well as imaging defects and doping profiles in Si or GaAs. Transient photoconductivity and optical rectification techniques of producing such pulses are outlined. The former techniques are also exploited in Chapter 5 for the investigation of charge transport in amorphous semiconductor image sensors. However, as the transit time scales for the two applications differ considerably, the instrumentation involved is very different. For example, femtosecond lasers are used as the light source in the THz applications, whereas for the study of image sensors, laser diodes or a xenon flash may be used.

Recent developments in amplitude, wavelength, phase and polarisation modulating sensors are described in Chapter 3. Amplitude modulating sensors and their applications are discussed in detail. Interferometric techniques employed in phase modulators and distributed optical fibre sensors are also covered. Readers interested in these particular sensors should find the list of further reading at the end of this chapter useful.

The subject of optical sensors is continued in Chapter 4, which deals with amorphous semiconductor photoreceptors and X-ray image sensors. This chapter opens with a review of the concepts and theories relevant to photoreceptors that are photoconductive devices employed in xerography. Fabrication, properties, characterisation and application of three classes of photoreceptors are then discussed: organic photoreceptors, semiconductor powder-resin binder layers and amorphous semiconductor photoreceptors. Several experimental techniques for the investigation of charge transport in these devices and typical experimental results are described. It should be noted that these techniques can be applied to other high resistivity materials and devices. Applications of amorphous semiconductors such as amorphous selenium and hydrogenated amorphous silicon for digital X-ray imaging are also included. This chapter concludes with a general coverage of recent advances in the optical application of amorphous semiconductors.

The polarisation and associated motion induced in otherwise uncharged particles by an ac electric field is known as 'dielectrophoresis'. In Chapter 5, the growing field of microbiological sensing by dielectrophoresis is reviewed, with an emphasis on the engineering aspects of instrumentation and transducer design. After describing the physical basis of dielectrophoresis, its application to sensing is discussed. Since electrodes (from which the ac driving field is applied) are important elements of any dielectrophoretic sensing system, their structure, arrangement and properties are described in some detail. The features of a computer controlled dielectrophoretic-sensing system using a CCD camera are also described and other examples of dielectrophoretic systems and measurement techniques are outlined. The chapter ends with a section on applications of dielectrophoresis, concentrating on reviewing a few recent studies of micro-organisms by this technique.

After a brief general introduction to compounds and their polymerisation, the synthesis of conducting polymers is described. Models used for describing the conduction process in such polymers are then reviewed. Conductivity modulation by absorption and desorption of volatile chemicals is discussed in some detail. Finally, array-based sensors and their response to volatile chemicals are described. The authors present experimental results to show that by normalising the steady state responses of the array sensors to different volatile chemicals or mixtures of chemicals, unique patterns are generated.

In chapter 6, organic semiconductors are broadly classified into the three classes of molecular crystals, polymers and charge transfer complexes. Molecular crystals are typified by materials such as anthracenes, naphthalene and phthalocyanines. The drastic change in electrical conductivity of thin film phthalocyanine when exposed to gases such as O_2 or NO_3 has provided the basis for an investigation of this material as a semiconductor gas sensor, which is the subject of Chapter 7. Although this chapter is focused on research into a particular thin film gas sensor, the techniques described are useful to researchers embarking on investigations of other thin films for similar applications. After a general introduction to ideal gas sensor qualities and different types of gas sensors, the rationale of thin film gas sensor characterisation is presented. This is followed by a brief description of experimental techniques for fabrication and investigation of dc and ac electrical, optical and gas sensing properties of thin film phthalocyanine (Pcs). Experimental results on the variation of dark current with differing temperatures in the presence of O_2 or NO_2 are described within the framework of a relevant theoretical model. Practical data comparing the sensitivity of these sensors to O_2 and NO_2 are also presented. The book concludes with a final chapter on thin film (ClAlPc) phthalocyanine gas sensors.

1.6 References

1 MORRIS, A. S.: 'The essence of measurement' (Prentice-Hall, 1996)
2 MORRIS, A. S.: 'Principles of measurement and instrumentation' (Prentice-Hall, 1993, 2nd edn.)
3 FIGLIOA, R. S., and BESLEY, D. E.: 'Theory and design for mechanical measurements' (John Wiley & Sons, 1995, 2nd edn.)
4 BENTLEY, J. P.: 'Principles of measurement systems' (Longman, 1997, 3rd edn.)
5 BOLTON, W.: 'Engineering instrumentation and control' (Butterworth Heinemann Ltd., 1996)
6 ROTH. P.: 'Ceramic PTC thermistors as sensors: Level measurement made easy', *Siemens Components Applications Products Trends*, 1998, **1**, p. 35
7 SINCLAIR, I. R.: 'Sensors and transducers' (Newnes, Oxford, 1992, 2nd edn.)
8 DTI, UK: *Sixth Sense*, Issue 21, Summer 1995, p. viii
9 HERNANDEZ, P. R., *et al.*: 'A new ion sensitive FET technology with back contacts using deep diffusion". Technical digest of the fifth international meeting on *Chemical sensors*, Pontificia Studiorum Universita, Rome, Italy, 11–14 July 1994, vol.1, **1**, p. 263
10 GRÁTTAN, K. T. V. (Ed.): 'Sensors technology, systems and applications' (Institute of Physics Publishing Ltd., UK, 1991)

11   ATKISON, J. K., SHAHI, S., VARNEY, M., and HILL, N.: 'A thick film electrochemical instrument', *Sensors and Actuators*, 1991, **B4**, p. 175

12   KRAUSE, S., MORITZ, W., and GROHMANN, I.: 'Improved long-term stability for LaF_3 based oxygen sensor', Proceedings of Eurosensors VII, Budapest, Hungary, *Sensors and Actuators* 1993, **B18**, p. 148

13   GRACIA, I., *et al.*: 'On-line determination of the degradation of ion sensitive FET chemical sensors', *Sensors and Actuators*, 1993, **B15–B16**

14   SCHEGGI, A. M., and BALDINI, F.: 'Chemical sensing with optical fibre', *International Journal of Optoelectronics*, 1993, **8**, p. 133

15   ROLT, K. D.: *Journal of the Acoustical Society of America*, 1990, **87**, (3), p. 1340

16   TERRY, S.: 'A miniature silicon accelerometer with built-in damping'. Technical digest of IEEE Solid State Sensor and Actuator Workshop, Hilton Head Island, SC, 1990, p. 44

17   SEIDEL, H., *et al.*: 'Capacitive silicon accelerometer with highly symmetrical design', *Sensors and Actuators*, 1990, **A21–A23**, p. 312

18   'Analogue devices combine micromachining and Bi CMOS', *Semiconductor International*, 1991, **17**, p. 17

19   FRENCH, P. J., and EVANS, G. R.: 'Polysilicon strain sensors using shear piezoresistance', *Sensors and Actuators*, 1988, **15**, p. 257

20   ANDREWS, M. K., *et al.*: 'A miniature resonating pressure sensor', *Microelectronics Journal*, 1993, **24**, (7), p. 831

21   'Novel sensors deployment techniques for marine environment', *Seasense News*, The Newsletter of the Seasense LINK Programme, Issue 2, Spring 1999, p. 1

22   'Flow meters', *Measurement and Control Journal*, 1998, **31**, (8), pp. 252–253

23   RADVANYI, A. G.: 'Structural analysis of stereogram for CNN depth detection', *IEEE Transactions on Circuits and Systems I: Fundamental Theory and Applications*, 1999, **46**, (2), p. 239

24   HOULT, S. R., and MIDDELHOEK, S.: 'High temperature silicon Hall sensor', *Sensors and Actuators*, 1993, **A37–A38**, p. 26

25   RICCOBENE, C., *et al.*: 'Operating principles of dual collector magnetotransistors studied by two-dimensional simulation', *IEEE Transactions on Electron Devices*, 1994, **41** p. 32

26   NOBLE, R.: 'A new direction in orientation', *Electronics World*, 1996, p. 40

27   BENNETT, A.: 'Hall-effect transducers in traction control', *Electrotechnology*, 1996, p. 16

28   PAUL, O., and BALTES, H.: 'Measuring thermogalvanomagnetic properties of polysilicon for the optimisation of CMOS sensors', Transducers 93 digest of technical papers, IEE, Japan, Tokohama, 1993, p. 606

29   VELCHEV, N.: 'Microelectronic Halltrons with high resistivity channel as effective semiconductor magnetosensitive structures', *Microelectronics Journal*, 1993, **24**, p. 639

30   SCHULIGA, A., *et al.*: 'Glucose-sensitive ENFET using potassium ferricyanide as an oxidising substrate, the effect of an additional lyscozyme membrane', *Sensors and Actuators B*, 1995, **24–25**, p. 117

31   BLUM, L. J., and COULT, P. R. (Eds): 'Biosensor principles and applications' (Marcel Dekker, New York, 1991)

32   HARDEMAN, S., *et al.*: 'Sensitivity of novel ultra thin platinum film

immunosensors to buffer ionic strength', *Sensors and Actuators B*, 1995, **24–25**, p. 98

33 RANK, M., *et al.*: *Applied Microbiology Biotechnology*, 1995, **42**, p. 813

34 SCHELLER, F., and SCHMID, R. D. (Eds), 'Biosensors: fundamentals, technologies and applications' (VCH, 1992)

35 ALBERGHINA, L., and SENSI, P. (Eds): Proceedings of the 6th European Congress on *Biotechnology* (Elsevier Science, 1994)

36 MURACK, A., and SCHUGER, K. (Eds): 'Computer applications in biotechnology'. CAB 95, DECHEMA, 1995

37 JURGENTS, H., *et al.*: *Analytica Chimica Acta*, 1995, **302**, p. 289

38 SCHUGERL, K., *et al.*: 'Challenges in integrating biosensors and FIA for on-line monitoring and control', a review paper, *Trends in Biotechnology (TIBTECH)*, 1996, **14**, (1), p. 21

39 *Proceedings of the IEEE*, 'Special issue on laser radar', 1996, **84**, (2)

40 LUTHON, F., and DRAGOMIRESCU, D.: 'A cellular analogue network for MRF-based video motion detection', *IEEE Transactions on Circuits and Systems I: Fundamental Theory and Applications*, 1999, **46**, (2), pp. 281–293

41 ROKS, E., *et al.*: 'A twin channel (p&n) FT-CCD image with cross-antiblooming', *IEEE Transactions of Electron Devices*, 1996, **43**, (2), p. 273

42 *Philips Journal of Research*, 'Special issue on solid state image sensors', 1994, **48**, (3), p. 145

43 YEE, H., *et al.*: 'Fabrication of high performance extended cavity double-quantum-well lasers with integrated passive sections', *IEE Proceedings – Optoelectronics*, 1996, **143**, (1), p. 94

44 CARLIN, D. B., and TSUNODA, Y.: 'Diode lasers for mass market applications: optical recording and printing', *Proceedings of the IEEE*, 1994, **82**, (4), p. 46

45 KONUMA, K. *et al.*: 'An infrared-bicolor Schottky-barrier CCD image sensor for precise thermal images', *IEEE Transactions on Electron Devices*, 1996, **43**, (2), p. 282

46 SHARMA, N. R., CAREY, W. P., and YEE, S. S.: 'Integrated electro-chemical microelectrode sensor array for gas and liquid detection'. Technical digest of the fifth international meeting on *Chemical sensors*, Rome, Italy, 1994, vol. 2, p. 114a

47 EGAN, A. *et al.*: 'Theoretical investigation of electro-optical synchronisation of self-pulsating laser diodes', *IEE Proceedings–Optoelectronics*, 1996, **143**, (1), p. 94

48 MEIJER, G. C. M., and VAN HERWAARDEN, A. W.: 'Thermal sensors', (Adam Hilger, Bristol, 1994)

49 'Engineering matters', Newsletter of the EPSRC Engineering Programme, March 99, Issue 1, p. 3

50 CAMERON, A. A.: 'Integrated night vision in helmet-mounted displays', *GEC Review*, 1999, **14**, (1), pp. 8–18

51 'A colour image capture board uses a hardware video scaler and a colour space converter', *Electronic Product Design*, 1999, **20**, (2), p. 62

52 SCHUPPERT, B., *et al.*: 'Integrated optics in silicon and silicon germanium heterostructures', *Journal of Lightwave Technology*, 1996, **14**, pp. 2311–2323

53 WAYNANT, R., and EDIGER, M. (Eds): 'Electro-optic handbook' (McGraw-Hill, New York, 1994)

54   GRIFFIN, P.: 'The photonics future', *GEC Review*, 1999, **14**, (1), pp. 3–7
55   WANG, C. C., *et al.*: 'Ultrafast all-silicon light modulator', *Optics Letters*, 1994, **19**, pp. 1453–1455
56   JAHAN, J., 'Optical computing hardware' (Academic Press, San Diego, USA, 1994)
57   HOMULA, J., and SLAVIK, R.: 'Fibre-optic sensor based on surface plasma resonance', *Electronics Letters*, 1996, **32**, (5), p. 480
58   RAO, B. K. N., HOPE, A. D., and WANG, Z.: 'Application of Walsh spectral analysis to milling tool wear monitoring', *Measurement & Control*, 1998, **31**, (7), p. 268
59   MacCRAITH, B. D., *et al.*: 'Development of a LED-based fibre optic oxygen sensor using a sol-gel derived coating', *Proceedings of SPIE*, 1994, **2293**, p. 110
60   THOMPSON, R. B., and LAKOWICZ, J. R.: 'Fibre-optic pH sensor based on phase fluorescence lifetimes', *Analytical Chemistry*, 1993, **65**, p. 853
61   'Fibre optic sensors for 3D surface measurement', *Electronic Product Design*, March 1999, p. C44
62   BOGUE, R., and Partners: 'The role of advanced sensors in environmental monitoring and pollution control practices'. A report on a study commissioned by the DTI, UK, November 1991
63   BOGUE, R. W., and Partners: 'A review of UK publicly funded research and development on advanced sensors'. Report for DTI, UK, November 1989
64   *Proceedings of the IEEE*, 'Special issue on laser radar', 1996, **84**, (2), p. 18
65   'Products for water industry', *Executive Engineer*, 1996, **4**, (1), p. 18
66   UENO, K.: 'Orientation dependence of the oxidation of SiC surfaces', *Physica Status Solidi (a)*, 1997, **162**, pp. 299–304
67   SPETZ, A. L., *et al.*: 'Current status of silicon carbide based high temperature gas sensors', *IEEE Transactions on Electron Devices*, 1999, **46**, pp. 561–566
68   HOLMES, C. B.: 'Desalination – the UK's answer to water demand?', *Executive Engineer*, 1996, **4**, (1), p. 12
69   PIETILA, P.: 'The role of engineers in the water sector is expanding', in VAEZI-NEJAD, S. M. (Ed.): Proceedings of Engineering Education Needs in Developing Countries Seminar, October 1995, London, UK. Sponsored by the IEE London Centre and SFEI, European Society for Engineering Education (University of Greenwich, March 1996), p. 32
70   OWEN, V. M.: 'Market requirements and opportunities for advanced sensors in the health care sector', Report for the DTI by Sci Tec., UK, November 1992
71   MOSELEY, P. T.: 'New trends and future prospects of thick and thin-film gas sensors', *Sensors and Actuators B*, 1991, **3**, p. 167
72   VELOS, C. O. G., SCHNELL, J., and CROSET, M.: 'Thin solid state electrochemical gas sensors', *Sensors and Actuators*, 1993, **2**, p. 371
73   ROH, Y. H., and LEE, B. H.: 'Thick film zinc oxide gas sensor for control of lean air/flow ratio in combustion systems'. Technical digest of the fifth international meeting on *Chemical sensors*, Rome, Italy, 1994, p. 792
74   Proceedings of Euro Sensors VII, 26–29 September 1993, Budapest, Hungary

75 WATTON, J.: 'Identification of fluid power component behaviour using dynamic flowrate measurement', Proceedings of the *IMechE*, 1995, **209**, p. 179

76 KATOH, K., *et al.*: Proceedings of the 13th Vienna International Motor Symposium, 1992, p. 249

77 FLACK, R. D., BRUNK, D., and SCHNIPKE, R. J.: 'Measurement and prediction of natural convection velocities in triangular enclosure', *International Journal of Heat and Fluid Flow*, 1996, **16**, (2), p. 106

78 WOLFF, H. H., *et al.*: 'Intelligent angular motion sensors', *Siemens Components Applications Products Trends*, 1998, **4**, p. 15

79 DOEBELIN, E. O.: 'Measurement systems application and design' (McGraw-Hill, New York, 1990, 4th edn.)

80 'Book of Abstracts', Eurosensors VII, Budapest, 1993, p. 15

81 ZUCKMANTEL, E., *et al.*: 'Optical engine oil level sensor'. Proceedings of the fifth international conference on *Automotive electronics*, London, IMechE, 1985, p. 325

82 Proceedings of the fifth international symposium on *Sensors and actuators*, 1993, Tranducers 93

83 MORGAN, C.: 'Lighting up the pressure points', *TAPM News*, 1998, (8), p. 6

84 FATAH, R. M. A.: 'Electrostatic activation of micromechanical resonators', *Electronics Letters*, **27**, (2), p. 16

85 MELDRUM, A.: 'Sensing the strain', *TAPM News*, 1998, (8), p. 3

86 BRAY, K. N. C., and COLLINGS, N.: 'Ionization sensors for internal combustion engine diagnostics', *Endeavour*, 1991, **16**, (1), p. 10

87 MOCK, R., and MEIXNER, H.: 'A miniaturised high temperature pressure sensor for the combustion chamber of a spark-ignition engine', *Sensors and Actuators A*, 1991, **25–27**, p. 103

88 GOTO, T., *et al.*: 'Toyota active control suspension system'. 22nd international symposium on *Automotive tech* (ISATA), Italy, 1990, p. 857

89 SUSUKI, S., *et al.*: 'Semiconductor capacitive-type accelerometers'. SAE technical paper 910274, 1991

90 TURNER, J. D.: 'The development of a thick-film non-contact shaft torque sensor for automotive applications', *Journal of Physics E*, 1989, **22**, p. 82

91 HAZELDEN, R. J.: 'Application of an optical torque sensor to a vehicle power steering system'. Proceedings of IEE Colloquium **C12**, May 1992, p. 107

92 TURNER, J. D., and PRETLOVE, A. J.: 'Acoustics for engineers' (Macmillan, London, 1991)

93 MORGAN, C.: 'Lighting up the pressure points', *TAPM News*, 1998, (8), p. 6

94 WEBSTER, J. G.: 'Electrical impedance tomography' (Adam Hilger, USA, 1996)

95 SEZE, R. D.: 'Epidemiology, human experiments and over exposures to assess health risk', *Measurement & Control*, 1998, **31**, (6), p. 176

96 ARMITAGE, A. F.: 'Eurosensors IX: The sensors of the future?', *Measurement & Control*, February/March 1996, **29**, p. 46

97 GROSSMANN, P.: 'Multisensor data fusion', *GEC Journal of Technology*, 1998, **15**, (1), p. 27

98   MAO, Y., and KRIER, A.: 'Efficient 4.2 μm LEDs for detecting CO_2 at room temperature', *Electronics Letters*, **32**, (5), 1996, p. 479
99   YU HUA, L. I., *et al.*: 'Multi-channel rotational speed measurement: a software based approach', *Measurement & Control*, 1998, **31**, (8), p. 229
100 DEARDEN, R.: 'Automatic X-ray inspection for food industry', *Electrotechnology*, 1996
101 'Sensing the future: Sensors in manufacturing and materials'. Report, Forsight Sensors Action Group, Department of Trade and Industry, October 1997, UK
102 AJLUNI, C.: 'Silicon accelerometer targets airbugs restraints systems', *Electronic Design*, 1995, **43**, (21)
103 MILT, L.: 'Micromachining technologies promise smarter sensors; actuators for a broad range of applications', *Electronic Design*, 1995, **43**, (25)
104 MEASURES, R. M.: 'Advances towards fibre optic based smart structurer', *Optical Engineering*, 1992, **31**, p. 34
105 GROVES-KIRKBY, C. J.: 'Optical-fibre strain sensing for structural health and load monitoring', *GEC Journal of Technology*, 1998, **15**, (1), p. 16
106 HERSEY, A. D., BERKOFF, T. A., and MOREY, W. W.: 'Fibre Fabry–Perot demodulator for Bragg grating strain sensors'. Proceedings of the ninth *Optical fibre sensors* conference, Florence, 1993, p. 39
107 FURSTENAU, N., JANZEN, D. D., and SCHMIDT, W.: 'Fibre optic interferometric strain gauge for smart structure applications: first flight tests'. AGARD conference proceedings 531, 'Smart structures for aircraft and spacecraft', 1993, p. 24–35
108 ESCOBAR, P., GUSMEROLI, V., MARTINELLI, M., LANCIANI, I., and MORABITO, P.: 'A fibre optic interferometric sensors for concrete structures'. Proceedings of the first European conference on *Smart structures and materials*, Glasgow, 1992, p. 215
109 BHATI, V., *et al.*: 'Application of absolute fibre optic sensors to smart materials and structures', Proceedings of tenth *Optical fibre sensors* conference, Glasgow, UK, 1994, p. 171
110 UTTAMCHANDANI, D.: 'Fibre-optic sensors and smart structures: developments and prospects', *Electronics & Communication Engineering Journal*, 1994, **6**, (5), p. 237
111 KIMURA, M., *et al.*: 'Application of the air bridge microheater to gas detection', Technical digest of the fifth international meeting on *Chemical sensors*, Rome, Italy, 1994, vol. 2, p. 1164
112 Proceedings of the ASCE Structure Congress XIII, Boston, 1995
113 WILLIAM, B., and SPILLMAN, J. R.: 'Sensing and processing for smart structures', *Proceedings of the IEEE*, 1996, **84**, (1), p. 68
114 WALSH, P. T.: 'Toxic gas sensing for the workplace', *Measurement and Control*, 1996, **29**, (1), p. 5

115 VAN EWYK, R.: 'Flammable gas sensors and sensing', *Measurement and Control*, 1996, **29**, (1), p. 13

116 PERSAUD, K. C., *et al.*: 'Measurement of sensory quality using electronic sensing systems', *Measurement and Control*, 1998, **29**, (1), p. 17

117 'European research on advanced sensors', Report, Hamer Associates sponsored by the DTI, UK, November 1990

118 VAEZI-NEJAD, S. M.: 'Primary investigation of anodised aluminium as a substrate for hybrid microelectronics I – Electrical and mechanical properties', *International Journal of Electronics*, 1990, **68**, (1), p. 59

119 CARSTENS, J. R.: 'Electrical sensors and transducers' (Prentice-Hall, Englewood Cliffs, NJ, USA, 1993)

120 Technical digest of the IEEE *Solid-state sensor and actuator* workshop, Hilton Head, SC, 1992, pp. 6, 72, 73, 109

121 VAEZI-NEJAD, S. M.: 'Primary investigation of anodised aluminium as a substrate for hybrid microelectronics II – Behaviour of thick film circuit components', *International Journal of Electronics*, 1990, **69**, (4), p. 519

122 CARLIN, D. B., and YOSHITO, T.: 'Diode laser for mass market applications – optical recording and printing', *Proceedings of the IEEE*, 1994, **82**, (4), p. 462

123 HUTCHESON, G. D., and HUTCHESON, J. D.: 'Technology and economics in the semiconductor industry', *Scientific American*, June 1996, **270**, p. 40

124 *Proceedings of the IEEE*, Special section on smart structures, 1996, **84**, (1), pp. 57–78

125 *Electronic Product Design*, April 99, p. 63

126 WEN-JENG, H. O., *et al.*: 'Highly uniform monolithic IX12 array of InGaAs photodiodes', *Electronics Letters*, 1996, **32**, (1), p. 61

127 See papers in Integrated Ferroelectrics, for example: SETIADI, T. D., *et al.*: 'A comparative study of integrated ferroelectric polymer pyroelectric sensors', *Integrated Ferroelectrics*, 1998, **18**, pp. 33–47

128 SEHODRI, S., *et al.*: 'Demonstration of an SiC neutron detector for high radiation environments', *IEEE Transactions on Electron Devices*, 1999, **46**, (3), pp. 567–571

129 VAEZI-NEJAD, S. M.: 'Inspection of materials for advanced sensors'. IEE Colloquium Digest No. 1995/232, pp. 14–20

130 ROTH. P.: 'Ceramic PTC thermistors as sensors: Level measurement made easy', *Siemens Components Applications Products Trends*, 1998, **1**, p. 35

131 DE CESARE, G.: 'Tunable photodetectors based on amorphous Si/SiC heterostructures', *IEEE Transactions on Electron Devices*, 1995, **4**, (5), p. 835

132 EBERHARDT, K., NEIDLINGER, T., and SCHUBERT, M. B.: 'Three color sensor based on amorphous n–i–p–i–n layer sequence', *IEEE Transactions on Electron Devices*, 1995, **42**, (10), p. 1763

1.7 Appendix: Some of the sensor and sensor related special interest groups, clubs and associations in the UK

Advanced Control Technology Club (ACT)	50 George Street, Glasgow G1 1QE
Advanced Control Technology Transfer Club	Laboratory of the Government Chemist, Queens Road, Teddington TW11 0LY
Diagnostic Club and Medical Sensors Special Interest Group (MSSIG)	The Diagnostics Club, Guild House, 31/32 Worple Road, London SW19 4EF
Effluent Processing Group	AEA Technology, 404 Harwell, Didcot, Oxfordshire OX11 0RA
Gas Sensors Group	Electrical Engineering Department, University of Wales, Swansea SA2 8PP
Industrial and Management Partnership in Automation and Control Technology (IMPACT)	50 George Street, Glasgow G1 1QE
Manufacturing and Machines Sensors Group (MMSG)	Elsym, PO Box 917, Newport Road, Cardiff CF2 2XH
M25 Club	University of Hertfordshire, Hatfield AL10 9AB
M62 Sensors and Control Club	AEA Technology, INSYT, Sales Office, Risley, Warrington, Cheshire WA3 6AT
Optical Sensors Collaborative Association (OSCA)	Sira Communications Ltd, South Hill, Chislehurst, Kent BR7 5EH
Process Tomography Group	Department of Electrical Engineering, UMIST, PO Box 88, Sackville Street, Manchester M60 2QD
Sampling Club	Laboratory of the Government Chemist, Queens Road, Teddington TW11 0LY
Scot Sense	Caledonian University, Cow Caddens Road, Glasgow G4 0BA
Sensors for Water Interest Group (SWIG)	27 West Green, Barrington, Cambridge CB2 5RZ
Ultrasonics and Acoustic Transducers Group (UATG)	'Greswoldes', Wedmans Lane, Rotherwick, Basingstoke RG22 9BT
United Kingdom Sensors Group (UKSG)	GAMBICA, Leicester House, 8 Leicester Street, London WC2 7BN

Chapter 2

Recent advances in measurement and instrumentation systems based on optical techniques

S. Hadjiloucas and D. A. Keating

2.1 General introduction to optical sensing

In the following section, a variety of optical sensors and techniques are described. However, the objective of this chapter is to make the reader aware of the vast possibilities and the recent developments on the subject from a systems based point of view only, rather than to shed light on the fundamental problems encountered by current research.

Optical measurements can generally be classified in the following categories:

- Imaging methods which are based on the recording of the spatial distribution of the light intensity, such as photography and microscopy.
- Interference methods based on the recording of the spatial distribution of the intensity and phase (energy and momentum) of the light waves, such as light interferometry and atom interferometry.
- Light scattering methods which are based on the measurement of the spatial and local features of scattered light using either single particle counters or performing particle concentration measurements based on diffraction, or rate of polarisation effects.

After a brief introduction to some commonly used coherent and incoherent sources, and some stabilisation techniques, a variety of detectors commonly used in sensor applications are described. Emphasis is then given to scattering techniques and the most commonly used spectroscopies, which very often may simultaneously incorporate a variety of modulation principles. Next, the different measurement techniques for the determination of electric dipole polarisabilities and some recent applications based on the interaction of light with matter, including the latest advances in THz measurement and instrumentation systems, are described. Finally, an introduction to optical fibre sensing is provided, as fibres are nowadays commonly used to accurately guide light to the location where measurements must be performed.

2.2 Introduction to optical sources

2.2.1 Broadband sources

Optical systems may be illuminated either by polychromatic (broadband) or by monochromatic sources. Of fundamental importance to an optical system illuminated by a broadband source is the number of modes propagating through it. This is the product of the source area and solid angle (or etendue, a parameter remaining constant throughout the system) divided by the square of the wavelength of the source. Therefore, although the radiant energy refers to the total energy emitted by a light source it is most common to refer to sources in terms of the radiance (W m^{-2} sterad^{-1}), which is the power emitted per unit surface element into a unit solid angle. Most isotropic sources are described by Lambert's law.

Another important parameter to sensor applications is the degree of temporal and spatial coherence of the source. The degree of temporal coherence may be observed from the fringe visibility function of a Michelson interferometer. The spatial coherence criterion in one dimension (coherence length) ensures that two partial amplitudes have a phase difference of less than 180 degrees, and for white light LEDs at 400–600 nm this is usually less than 1 mm.

Extension of this coherence condition in two dimensions defines the maximum surface area that can be coherently illuminated, and this in turn defines the maximum solid angle inside which the radiation field shows spatial coherence. For a plane wave produced at the focus of a lens, spatial coherence is observed over the whole aperture confining the light beam, and the coherence surface increases with the square of the distance from the source. Using a Gaussian–Schell source model, it is possible to identify the number of coherent modes needed to describe a field so that a coherent representation of partially coherent beams can be performed [1].

Multiplying the coherence length of a source in the propagation direction by the coherence surface area produces a coherence volume that contains a mean number of photons within a particular spectral range. The mean number of photons per mode is often called the degeneracy parameter of the radiation field. This is identical to the mean number of photons per mode of the thermal radiation of a source predicted through Planck's law.

The equivalent broadband source at the far-infra-red part of the spectrum is the mercury-arc lamp, and sources at even longer wavelengths are possible using reverse biased diodes to avalanche breakdown mounted on a linearly tapered slot antenna [2].

2.2.2 *Intensity stabilisation technique for modulated LED sources*

When LEDs are used as light sources in sensor applications, the technique of amplitude modulation and phase sensitive detection is commonly used to reduce ambient noise. Furthermore, in order to improve the thermal stability of the source which is temperature and time dependent, amplitude modulation should be used in conjunction with intensity referencing. A simple way of achieving this is through a local feedback path that stabilises the LED source. The forward path of the circuit in Figure 2.1 comprises a differential amplifier, a high gain amplification stage and a voltage to current converter (V–I). The feedback path contains a photodiode, a current-to-voltage converter (I–V), a phase sensitive detector (PSD) with low pass filtering, and a compensator. When applying negative feedback, for high loop gains the system response is governed by the characteristics of the feedback path, and the non-linearities in the forward path that drive the LED are reduced by a factor of (1 + loop gain).

Phase sensitive detectors (PSD) are in common use in physical instrumentation where accurate measurement of a small ac signal is obscured by noise [3–6]. Generally a PSD produces a dc output signal proportional to the amplitude of an ac signal at a desired frequency that is equal to the frequency of the ac reference signal (synchronous detection). A basic property of a PSD is that it rectifies the signals

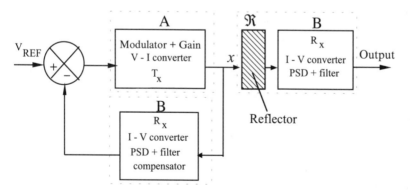

Figure 2.1   *Block diagram showing the local feedback for thermal stabilisation of the emitter. For large forward gain A, and provided that the two receiving circuits with overall transfer function B are matched,*
$$Q \cong (1/B) \, \Re B = \Re$$

(positively for antiphase signals and negatively for in-phase signals). Although rectification and low pass filtering of an amplitude modulated signal can be used to shift the signal to dc level, a PSD is preferred as it offers a 3 dB advantage over a rectifier. A further useful property of a PSD is that the output is proportional to the cosine of the phase angle between the input and the reference signal. The polarity dependence of the output makes the PSD a useful sign sensor (a particularly useful property when the PSD is employed in a feedback system).

A PSD can be regarded as a multiplier followed by a low pass filter where the product of the signal and the reference (typically, a square wave), can be represented as a harmonic series from which the dc contribution is extracted as the higher frequencies are attenuated by the low pass filter. Therefore a dc output is produced only when the signal frequency and the reference are identical and phase coherent. Signals that are phase coherent but in quadrature with the reference give zero output and this quadrature rejection property provides an improvement in detection. The overall response of the PSD is therefore equivalent to a narrow bandwidth band pass filter centred at the reference frequency. The effective bandwidth is determined by the half power bandwidth of the low pass filter and the only noise that passes through it is that occurring in the bandwidth of $1/4\tau$ (where τ is the time constant of the filter used). As τ increases, the bandwidth decreases, and the input acceptance of the PSD is narrowed, rejecting most of the input noise so that the signal recovery is improved. There is a limit to the improvement that can be achieved for signal recovery as further increasing the time constant of the filter of the PSD for a signal with a high harmonic content will result in a severe distortion of the waveform. Furthermore, there is a limitation to the amount of noise a PSD may accept, and if a very large bandwidth of noise is used, the PSD may be overloaded. Finally, there is an advantage in incorporating a PSD into a system, as the signal-to-noise ratio can be further improved by selecting the reference signal in a region where it can be amplified with minimal amplification of $1/f$ noise (of importance when using detectors with a very low noise floor).

A very simple switching PSD can be realised when the reference signal from a stable oscillator is fed to the non-inverting input of an op-amp and the output is the result of the multiplication of the sensor signal with the reference. By ensuring unit gain in the op-amp, its output is switched with the aid of an analogue switch between ± 1, in phase with the reference square wave. Because the PSD does not have to produce a continuously variable gain, the active devices in the reference path can work either at saturation or at cut-off and hence a large dynamic range of operation is possible. The circuit described in the article by Irvine *et al.* [7] has been developed to improve the operation of the PSD under large amplitude interfering signals and also to minimise the number of components and the setting up procedures. The switching action of the reference signal can simply be controlled by a 4069 CMOS hex inverter chip. In a closed loop arrangement, an odd number of inverters will always oscillate (Figure 2.2). The frequency of oscillations is

$$f = \frac{1}{2C(0.405R_{eq} + 0.693R_1)} \tag{1}$$

where $R_{eq} = R_1 R_2 / (R_1 + R_2)$. A 4013B D-type clocked flip-flop can be used as a divide-by-two counter to ensure an equal mark-to-space ratio of the square wave oscillator.

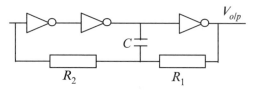

Figure 2.2   Circuit diagram of the square wave oscillator where the output voltage is controlled by different gate states

The signal from the transimpedance amplifier may be fed to the op-amp which switches from an inverting to a non-inverting configuration through the switching of the REF signal provided by a switch multi-plexer. When the REF signal is high, the circuit is equivalent to the one in Figure 2.3a, with an input resistance of $R/3$, and the op-amp has a unit gain. When the REF signal is low, the circuit is equivalent to the one in Figure 2.3b, the equivalent input resistance is $R//(R/2)$, and the output from the op-amp is -1 (as long as the ON resistance of the switches is in the region of a few ohms). The analogue switches are so arranged that the unused input of the amplifier is always switched to earth to maintain equal source resistance at the input of the amplifier. Using the switch connecting $R/2$ to earth keeps the resistance seen by the two inputs approximately equal. Resistor matching at the input stages of the op-amp ensures a small offset voltage due to a good common-mode rejection ratio. Although the symmetry of the circuit minimises the resulting output offset, the use of a single integrated circuit for all the switches

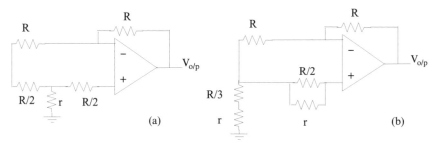

Figure 2.3   Equivalent circuits of the (a) inverting and (b) non-inverting configurations in the PSD

would ensure close matching characteristics. The switches may all be contained in a single integrated CMOS 4066B multiplexer chip.

The linearity of the PSD is dependent on the variation of the ON resistance of the switches with signal voltage and temperature, and also on the magnitude of the transient spikes. The effect of the ON resistance variation can be minimised by keeping the resistors in the circuit as high as possible. This has the added advantage of minimising the magnitude of the transient spikes. However, the greater the values of these resistors, the worse is the noise performance of the amplifier. It is important that the amplification of the modulated ac signal from the transimpedance amplifier before the PSD is such that the equivalent input noise voltage will not be degraded by the noise level of the PSD.

In order to ensure stability under closed loop operation of the system in Figure 2.3, measurements of gain and phase must be performed at different frequencies. The resulting gain and phase Bode plots of the overall loop can then be calculated after taking into account the optical coupling of the system. Figures 2.4a and 2.4b show that a forward gain of 4000 is possible with such a system. It can be seen that the phase margin of the loop is 6 degrees and the gain margin is 26 dB. Although the phase margin is relatively small, the system is stable.

When applying negative feedback to a system, for high loop gains the system response is governed by the characteristics of the feedback path B. Therefore, for a very linear feedback characteristic, the non-linearities in the forward path A will be eliminated, and using transfer function notation for Figure 2.1:

$$Q = \frac{OUTPUT}{REF} = \frac{A}{1 + AB} \cong \frac{1}{B} \qquad (2)$$

From this analysis, it can be observed that, using a closed loop configuration, the response of the voltage-to-current converter is not critical for good thermal stability of the emitter. In addition, the sensitivity of the control system used to stabilise the LED to parameter variations can be calculated. Defining the system sensitivity as the ratio of the percentage change in the system transfer function Q, the sensitivity S becomes

$$S = \frac{\Delta Q/Q}{\Delta A/A} \cong \frac{\partial \ln Q}{\partial \ln A} \qquad (3)$$

The sensitivity of the system to changes in the forward path A is

$$S_A^Q = \frac{\partial Q}{\partial H} \frac{A}{Q} = \frac{1}{(1 + AB)^2} \frac{A}{A/(1 + AB)} = \frac{1}{(1 + AB)} \qquad (4)$$

and the sensitivity to changes in the feedback element is

$$S_B^Q = \frac{\partial Q}{\partial H} \frac{H}{Q} = \left[\frac{A}{1 + AB} \right]^2 \frac{-B}{A/(1 + AB)} = \frac{AB}{1 + AB} \cong -1 \qquad (5)$$

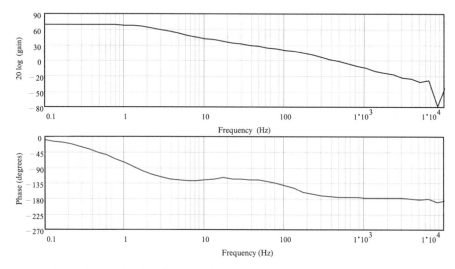

Figure 2.4 (a) Bode plot showing the gain and (b) phase for the overall feedback loop

Finally, in order to improve further the performance of the overall circuit with respect to noise, the bias current from the transimpedance amplifier in the forward path (corresponding to the standing current xI_a at the mid-point of the linear region of the operational range of the transducer), must be matched with the standing noise current I_b from the transimpedance amplifier in the feedback path using a custom-made optical coupler. The design of the optical coupler therefore requires that $1 = xI_a + I_b$ and $I_a = I_b$ so that $I_b = 1/(1 + x)$.

2.2.3 *Lasers*

Lasers generally consist of an optical resonator, an active medium which amplifies an incident electromagnetic wave and an energy pump which selectively pumps energy into the active medium so as to achieve population inversion through positive feedback. Depending on the condition under which lasing is achieved lasers are divided into semiconductor diode, Raman, tuneable solid state, colour centre, dye, excimer, free-electron and X-ray lasers.

Furthermore, lasers may operate under either a continuous wave (CW) or a pulsed mode. From a historical point of view, after the realisation of the Nd glass laser, pulsed dye lasers and CW dye lasers appeared. Active mode-locking using either an acousto-optic modulator or a Pockels cell for a CW He–Ne laser have also been used to produce pulses with durations of the order of 500 ps, with typical pulse energies in the region of

0.1 nJ. This was improved in later years using passive mode locking which produced pulses with a duration of 1 ps.

The development of colliding pulse mode-locked lasers, where a ring dye laser with two counter-propagating pulses travelling in an absorbing medium is passively mode locked by collision, has led to pulses with durations of less than 100 fs. The idea of self-phase modulation in an optical fibre, where the linear dispersion of the fibre results in a pulse spatial broadening and the intensity dependent refractive index component results in a spectral broadening of the pulses, has produced further advances in the field. Once further compressed using a grating pair, these pulses have durations of less than 10 fs.

Most recently, it has been realised that a suitable choice of pulse intensity can nullify the dispersion effects caused by the linear dispersion mechanism and therefore pulses are capable of propagating inside a medium without changing their temporal profile. Such soliton pulses are capable of producing stable femtosecond pulses in a Ti : sapphire laser. The pulse duration in these pulsed lasers is conveniently measured using either streak cameras or an optical correlation technique.

These recent advances in fast pulse devices have allowed developments in time resolved spectroscopies which are of fundamental importance to many branches of chemistry, physics and biology. Electronic transitions in the ultraviolet and visible parts of the spectrum have typical lifetimes of the order of 10^{-9} and 10^{-7} s, respectively, whereas the vibrational transitions in the near- and mid-infra-red regions have lifetimes between 10^{-7} and 10^{-1} s. The only slow transitions are the rotational ones in the far-infra-red and microwave parts of the spectrum with lifetimes between 10^{-1} and 10^{3} s.

It is interesting to note that the potential of fast pulse techniques in imaging has not been fully realised yet as the broad spectrum of a single narrowly focused pulse in a small object contains frequencies below the diffraction limit imposed by the object itself. Although the spectral resolution of most time-resolved techniques is in principle confined by the Fourier limit, it is possible to use regular trains of short pulses, thus circumventing the Fourier limit of a single pulse and reaching extremely high spectral and time resolutions.

In addition to the ability to produce fast pulses, another attractive feature of lasers for measurement purposes is their long coherence length. Typically, a multimode ion laser at 515 nm has a spectral bandwidth of 10 GHz and a coherence length of 1 cm, whereas a single-mode diode laser at 780 nm has a 50 MHz bandwidth and a coherence length of 2 m. Although a good single-mode He–Ne laser at 633 nm can reach a coherence length of about 100 m, certain measurement applications require the use of actively stabilised narrow linewidth He–Ne lasers which can reach a spectral bandwidth of only 50 kHz and a coherence length of 2 km.

2.2.4 *Intensity and wavelength stabilisation and tuning techniques for coherent sources*

The intensity of a CW laser is not completely constant but shows periodic and random fluctuations as well as long term drifts. Gas lasers are particularly vulnerable to power supply ripple which can occasionally lead to a modulation of their intensity. Other sources of noise are instabilities of the gas discharge, dust particles in the resonator, vibrations of the resonator mirrors and, in multimode lasers, mode competition. In CW dye lasers, density fluctuations in the dye jet stream and air bubbles are the main cause of intensity fluctuations.

For gas lasers, long term drifts of the laser intensity are generally caused by temperature and pressure fluctuations in the gas discharge, by thermal de-tuning of the resonator, and by degradation of the optical quality of the mirrors, windows and other optical components in the resonator. These intensity fluctuations lead to a lower signal-to-noise ratio of the sensing technique and therefore a variety of methods exist in the literature that describe the stabilisation of laser intensity.

It is possible to stabilise the intensity of a laser source by splitting a fraction of the output power with a beam splitter to a detector and comparing the voltage output of the detector to a reference voltage so that the difference is amplified and fed to the power supply of the laser, thus controlling the discharge current. This loop is suitable over the region in which the laser intensity increases linearly with current, but the technique is suitable only for small frequencies of modulation.

To compensate for intensity fluctuations in the MHz range, the output of the laser can be sent through a Pockels cell, which consists of an optically anisotropic crystal placed between two linear polarisers. An external voltage applied to the electrodes of the crystal causes optical birefringence and rotates the polarisation plane, changing the transmittance through the second polariser. Again, by detecting part of the transmitted light, the voltage from the photodetector can be compared to a voltage reference, and the difference can be used to drive the feedback control circuit (usually a PID controller). Because one has to bias the Pockels cell to work on the slope of the transmission curve, an intensity loss of 20% to 50% of the laser output results by the use of this technique.

For some other applications, it is essential that the laser wavelength stays as stable as possible at a pre-selected value during the measurement period. When spectral purity is important, single-mode lasers are generally used. Since the laser frequency is directly related to the laser wavelength, the technique is often quoted in the literature as (optical) frequency stabilisation, although for most methods in the visible spectral region it is not the frequency but the wavelength which is directly measured and compared with a reference standard. In the infra-red region however, most wavelength stabilisation methods rely directly on absolute frequency measurements instead.

Similarly to intensity, long-term frequency drifts and short-term fluctuations can degrade the spectral purity of laser sources. Long-term fluctuations are very often temperature-linked and therefore low thermal expansion materials must be used (Invar, fused quartz, Zerodur). In many designs, however, these drifts are compensated by electronic servo-control. The long-term fluctuations are mainly due to acoustical vibrations of the resonator mirrors and therefore an optical table is commonly used. In the absence of an optical table, an acoustically isolated table with sorbothane legs (for acoustic damping), a styropor matrix and a sand layer, that damps the resonances of a granite block resting on it, may be used instead.

Most wavelength stabilisation systems consist of three main elements: a wavelength reference standard with which the laser wavelength is compared, a controller system which for gas lasers tunes the resonator length, and an electronic feedback control element that tries to minimise the deviation of the laser wavelength from its reference value. Detecting the wavelength at the maximum or on the slope of a transmission peak of a Fabry–Perot interferometer maintained in a controlled environment can be used as a reference. Alternatively, the wavelength of an atomic or molecular transition (such as CH_4 or an inert gas) may serve as a reference.

Occasionally, another stabilised laser (such as a He–Ne laser) can be used as standard and the laser wavelength is locked to this. The residual frequency fluctuations of a stabilised laser are usually presented in an Allen plot, in which the *Allen variance* [8], which is a measure of the relative standard deviation in frequency difference between two lasers stabilised on to the same frequency and measured several times at equal time intervals, is used. More complete surveys of wavelength stabilisation can be found in the articles by Tomlinson and Forks [9] and Hough *et al.* [10]. An in-depth report on the frequency stabilisation of semiconductor laser diodes can be found in the book by Ikegami *et al.* [11].

In some applications, tuning of a single-mode laser continuously through a small range of wavelengths is desirable. This can be easily achieved by changing the optical path length between the resonator mirrors using a linear voltage ramp and a piezo-element. However, the limited tuning range and the hysteresis in the expansion of the piezo-element restrict its use to a small range of applications. Larger tuning ranges can be achieved by tilting a plane-parallel glass plate around the Brewster angle inside the resonator. In order to avoid a translational shift of the laser beam when tilting the plate, two plates simultaneously tilted in opposite directions can be used. This technique can further be used to modulate the wavelength of a single mode dye laser by tuning it to the reference wavelength of a tuned reference Fabry–Perot cavity. Alternative wavelength tuning techniques include the use of lasers with a broad gain profile, the use of techniques that shift the energy levels in the active medium (such as the Zeeman effect) or the use of optical frequency mixing.

For semiconductor laser diodes the spectral range of spontaneous emissions can be varied within wide limits by appropriate selection of semiconductor materials and their composition as overlaying blocks. For certain resonator modes within the spectral gain profile, temperature tuning changes the bandgap energy of the semiconductor materials leading to mode hops in a wide frequency range. At the same time, changing the current supplied to the diode provides a continuous tuning in a narrow range around a single mode. Most recently [12], a thermal equivalent circuit of a laser diode has been produced directly relating temperature noise to frequency noise.

In tuneable solid state lasers, the absorption and emission spectra can be varied within a wide spectral range by doping them with atomic or molecular ions, which broaden and shift the ionic energy levels of the lattice. The absorption spectra of such lasers depend strongly on the polarisation direction of the pump light, and therefore continuous tuning over their spectral gain profile is possible.

2.3 Overview of detection schemes

The most widely used detector in amplitude, phase and wavelength modulating sensors is the photodiode. The ideal photodiode can be considered as a current source, in parallel with a semiconductor diode. The current source corresponds to the current flow caused by a light generated drift current, while the diode represents the behaviour of the junction in the absence of incident light. The standard diode equation is

$$I = I_{photo} + I_{dk}(e^{qV_0/kT} - 1) \tag{6}$$

where I is the total device current, I_{photo} is the photocurrent (a reverse current), I_{dk} is the dark current, V_0 is the voltage across the diode junction, q is the modulus of the electron charge, k is Boltzmann's constant, and T is the temperature (in K). The photogenerated current is additive to the diode current, and the dark current is the diode reverse leakage current. The detector shunt resistance is the slope of the I–V curve, dI/dV evaluated at $V = 0$.

Since photodiodes are quantum devices; each incoming photon will generate one or zero units of electron charge which will contribute to the photocurrent. The probability of generating a charge is termed the quantum efficiency. Quantum efficiency mainly depends on how efficiently charge carriers are swept across the junction. The responsivity $\Re$ quantifies the photo-electric gain of a detector. The photodiode responsivity is the ratio of the photocurrent generated for each watt of incident light power (A W^{-1}). The responsivity depends directly on the quantum efficiency. The energy carried by each photon depends on its wavelength, $E = h\omega$, where ω is the photon frequency and h is Planck's constant (note that longer wavelengths in the far-infra-red part of the spectrum carry less energy). Therefore, expressing the responsivity in amps watt^{-1} gives this parameter an inherent wavelength dependency

$$\Re = \frac{q\eta\lambda}{hc} \tag{7}$$

where c is the speed of light and λ the wavelength of the detected photons, and η is the quantum efficiency of the detector. Responsivity has an additional wavelength dependency arising from the variation of quantum efficiency with wavelength. At wavelengths where silicon does not absorb strongly, photons may penetrate more deeply into the device leading to a minority carrier generation too remote from the junction to be detected, resulting in a lower quantum efficiency. The typical shape of the silicon photodiode responsivity spectral curve is determined by the absorption spectrum of silicon. Therefore photodiode responsivity should be specified at one wavelength unless a complete wavelength calibration is performed.

The responsivity of a photodiode remains constant even at low values of device saturation (low values of illumination) and therefore the device can be assumed to be linear. System linearity is also affected by the sensing circuit. Incident light falling on the photodiode active area produces a photocurrent which can be measured by the amount of voltage drop across an external resistance of known size, or better, using a current amplifier. As the resultant voltage in the sensing circuit increases, the photodiode becomes forward biased, leading to non-linear response.

When designing a sensing circuit to maximise the speed or linearity of response, it is essential to know two important electrical characteristics of the photodiode: the junction capacitance C_j and the shunt resistance R_{sh}. Without these, the RC time constant of the complete operating circuit (Figure 2.5) cannot be calculated.

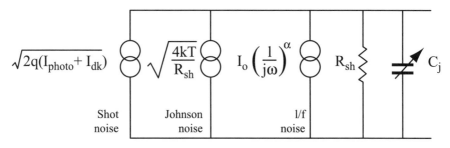

Figure 2.5   Noise equivalent circuit for a photodiode

The most significant figure of merit for the photodetector is the noise equivalent power (NEP). This is defined as the incident light level radiated on a photodiode that produces a photocurrent equal to the noise level. The NEP (typically of the order of 10^{-14} W Hz$^{-1/2}$ for most well designed systems), is a function of the photodiode responsivity, the noise of the photodiode and the associated sensing circuit, as well as the frequency bandwidth over which the noise is measured. The signal-to-noise ratio (SNR) may be computed by taking the ratio of the incident optical power to the optical power of the photodiode NEP$_{rms}$ = $I_{rms}/\Re$ (where $\Re$ is in amps watt^{-1}). The RMS noise current is the total integrated noise over the frequencies of interest:

$$I_{rms} = \sqrt{\int_{f_1}^{f_2} \left[\sum_{i=1}^{N} i^2(f) \right] df} \qquad (8)$$

If the bandwidth over which the noise is integrated is 1 Hz, at frequency f this is referred to as the noise spectral density (watts Hz$^{-1/2}$). The detectivity D of the photodiode is inversely proportional to the NEP and is normalised to the area of the detector (cm Hz$^{1/2}$ watt^{-1}).

The lower limit of the photodiode is determined by the Johnson noise, the shot noise and the $1/f$ noise. For modulated signals in the kHz range however, the $1/f$ noise is negligible. Furthermore, some corrections should be made for the dark current and the shunt resistance of the photodiode to compensate for the operational temperature (above 25°C).

There is a variety of possible circuits that may be used with photodiodes. A resistive load circuit where the photodiode is not reverse biased is very simple in operation but suffers from limited speed, linearity and sensitivity. Reverse biasing the photodiode extends the linearity and improves the speed but the shot noise increases due to an increase in the dark current. The use of a transimpedance amplifier, without reverse biasing the photodiode, further improves the linearity and speed and reduces the noise from the photodiode, but the additional amplifier noise must be taken into account in the overall noise performance calculations. Finally, in applications where maximum speed and linearity are desirable, a transimpedance amplifier with the photodiode reverse biased may be used as long as the increased shot noise and dark current offset are not so important in the intended application.

When using uncompensated op-amps a resistor and a capacitor in series may be added between the two inputs to stabilise it (Figure 2.6a). This offers the advantages of high speed and linearity without introducing excess shot noise. If a bipolar output is required, the alternative configuration in Figure 2.6b may be used.

The dominant noise source in the system is the shot noise of the output photodiode. This is given by $i_n = (2eBI)^{1/2}$ where e is the electronic

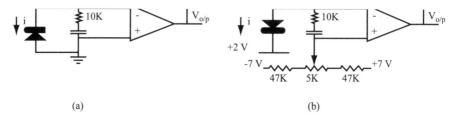

Figure 2.6   (a) Unipolar and (b) bipolar receiver circuits

charge, I is the diode current, B is the bandwidth and i_n is the rms noise current. Furthermore, the thermal noise should also be taken into consideration, using $i_t = (4kTB/R_f)^{1/2}$, where k is Boltzmann's constant, T is the absolute temperature, and R_f is the feedback resistor in the current-to-voltage converter. The signal-to-noise ratio (SNR) is then defined as the standing current at the operating point (mid-point of the linear region) divided by the combined noise currents $\sqrt{i_n^2 + i_t^2}$.

There is a plethora of techniques that may be used to optimise the performance of a photodiode for a particular sensing application. A typical example is the use of a dual diode preamplifier for laser Doppler interference measurements as shown in Figure 2.7. Using this configuration [13] the current due to the surrounding light incident on both diodes is eliminated.

In most circuits, it is preferable that FET op-amps are used, because of their high input impedance and their low input current noise. In some cases, a capacitor may be added in parallel to the feedback resistor to prevent oscillations at high signal levels. This results, however, in a limited frequency response of the circuit. Excellent reviews on preamplifier design using photodiodes may be found in the articles by Eppeldauer and Hardis [14] and Fonck *et al.* [15].

Alternatively, there is a variety of other detectors whose operational principles are very similar to that of photodiodes, and are therefore described below. Avalanche diodes are reverse-biased semiconductor diodes that use an internal amplification of the photocurrent. Their advantage over photodiodes is their fast response time and their high responsivity. For sufficiently high breakdown voltages, the gain bandwidth product of avalanche diodes is 10^{12} Hz. Their disadvantage over photomultipliers is the small active area, though this makes them suitable for optical fibre sensing applications.

By integrating a large number of small photodiodes on a single chip, a photodiode array can be made. A typical 1024 diode array is 22 μm long and 40 μm high. Every photodiode is connected by a multiplexing MOS switch to a voltage line and is recharged to its original bias voltage with recharging pulses, the pulses being a measure of incident radiation energy. Peltier-cooling the array significantly reduces the dark current

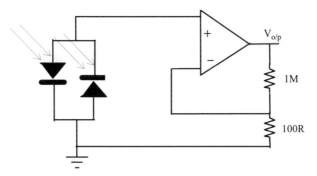

Figure 2.7   Dual diode preamplifier for detection of laser Doppler interference photons

and allows for fast integration times. At room temperatures, however, typical sensitivity limits are about 5000 photons per second.

Diode arrays are particularly useful as sensing elements of spectrum analysers. If a linear diode array of length L is placed on the observation plane of a spectrograph, the spectral interval $\delta\lambda = Ld\lambda/dx$ depends on the linear dispersion $dx/d\lambda$ of the spectrograph. The smallest resolvable spectral interval is then $\Delta\lambda = bd\lambda/dx$, and is limited by the width b of a single diode. Fitting a spectrometer with such an array, so that the path length difference is dictated by the grazing angle of the elements in the array, eliminates the need for a travelling mirror, leading to an instrument with improved mechanical stability and an improvement in noise due to the parallel interrogation rather than the sequential one which is commonly used in Fourier transform spectrometry.

Photodiode arrays are now frequently replaced by charge-coupled device (CCD) arrays which consist of an array of MOS junctions on a doped silicon substrate. Upon illumination, the charge of the MOS changes, leading to a change of its capacitance. A sequence of voltage steps shifts these changes of charge to the next diode, until the final diode is reached. The major advantage of CCD arrays is their large spectral range (100–1000 nm) and their large dynamic range, which covers about five orders of magnitude.

Another important type of detector is the photo-emissive detector. This device is made of materials that have a low work function and, upon illumination with monochromatic light, emits photo-electrons which are further accelerated by a voltage between an anode and a cathode. The resultant photo-current is measured either directly or by the voltage drop across a resistor. Detectors that rely on the photo-emissive principle are photocathodes, photocells, photomultipliers and photo-electric image intensifiers.

Photocathodes may have two types of photo-electron emitters in their cathode, an opaque layer or semitransparent layers. The quantum efficiency of the devices depends on the cathode material, and their spectral sensitivity extends to 1.2 μm. However, using new photocathodes based on photoconductive semiconductors whose surface has been treated to obtain a state of negative electron affinity, a high sensitivity over an extended spectral range can be achieved. In photocells, the photocathode is generally of the reflection type and the response time is of the order of few picoseconds. In most applications now, however, fast photodiodes are replacing them. For the detection of low light levels, photomultipliers are better suited than photocells.

Photomultipliers overcome the problems of noise limitations by internal amplification of the photocurrent using secondary-electron emission from internal dynodes that multiply the number of photoelectrons. However, these devices in the presence of a short duration light pulse produce a time spread voltage pulse at their output, as initial velocities of the emitted photo-electrons arriving at the dynodes will differ. Furthermore, the time of flight between the cathode and the first dynode strongly depends on the location of the spot on the cathode where the photo-electrons are emitted. This results in a further time spread of the output, which can only be avoided by focusing the incident light beam. A common technique further to reduce the rise time of an anode pulse is to decrease the operation voltage. Because of the need for dedicated power supplies and high operational voltages, photomultiplier based systems have been generally bulky and expensive in the past, although this is not so much the case at present (i.e. Hamamatsu). Their internal amplification mechanism has made them extremely suitable for single-photon counting techniques where a fast discriminator is normally used to provide the signal to the counter.

Another type of photo-emissive detector is known as a photo-electric image intensifier. It consists of a photocathode, an electro-optical imaging device and a fluorescence screen in which the radiation pattern is intensified by accelerated photo-electrons. With appropriate gating these devices can be used for detection of signals with high time resolution, although due to their high spatial resolution their major application is in spectroscopy.

Finally, another important type of detector is the thermal detector. Because of their wavelength independent sensitivity, thermal detectors are useful for calibration measurements (absolute measurements). A model of a thermal detector is shown in Figure 2.8. For an incident power P_0 the responsivity S of a thermal detector with heat capacity H and thermal conductivity G at a temperature difference ΔT from its environment is $S = \Delta T / P_0$. For large responsivities, G and H must be made as small as possible. At high modulation frequencies of incident radiation the time constant $\tau = H/G$ limits the frequency response of the detector, and therefore a fast and sensitive detector should have a minimum heat capacity. Generally, simple calorimeters fitted with

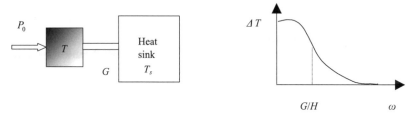

Figure 2.8   *Model of a thermal detector. Typical G/H corner frequencies for pyroelectric detectors are of the order of 20 Hz, whereas for a liquid He cooled cyclotron enhanced hot electron bolometer they are above 1 MHz*

thermocouples can be used as detectors, although configurations using a null balance technique within the calorimeter or externally with a balanced bridge circuit are preferred.

For more sensitive detection of low incident powers, bolometers and Golay cells are used. Bolometers have been used for both the infrared and the far-infra-red parts of the spectrum, and consist of many thermocouples in series, where one junction touches the reverse side of a thin electrically insulating foil exposed to the incident radiation, while the other junction is in contact with a heat sink. In the case of a superconducting bolometer, a feedback circuit is used to maintain the thermocouples at their critical superconducting temperature. The incident radiation power can be very sensitively measured from the magnitude of the feedback control signal used to compensate for the absorbed radiation power. Although these detectors when cooled offer NEP values in the region of 10^{-9} to 10^{-12} W Hz$^{-1/2}$, a new class of commercially available cyclotron enhanced hot electron detectors cooled at liquid helium temperatures (e.g. QMC Instruments, UK) can provide performances of the order of 10^{-13} W Hz$^{-1/2}$ in the far-infra-red part of the spectrum. State-of-the-art web-type designs offering minimum heat capacity and superior thermal conductivity characteristics have been reported to provide performances of the order of 10^{-18} W Hz$^{-1/2}$.

In the Golay cell detector (which may be viewed as an optical microphone), the detected radiation expands the air in an enclosed capsule, forcing a flexible membrane, on which a mirror is mounted, to move. Normally, the movement of the mirror is monitored by observing the deflection of a light beam from a light emitting diode, although in some designs, the flexible membrane forms part of a capacitor plate (similar to a conventional microphone).

More recently, pyroelectric detectors can be used as they are more robust than Golay cells, and therefore less delicate to handle. Their operation relies on the high sensitivity of their macroscopic electric-dipole moment with temperature. A change of the internal polarisation caused by a temperature rise produces a measurable change in surface charge, which can be monitored by a pair of electrodes applied to the element.

Although in the far-infra-red part of the spectrum absolute power detection in the past has been performed using a thermopile radiometer modified from the visible region, more recently, power detection for electromagnetic waves travelling in free space is performed using a photo-acoustic power meter (Thomas Keating Ltd, UK). The photo-acoustic power meter comprises a nichrome film on a thin Mylar substrate enclosed within a gas cell formed by the space between a transparent window and an adjustable back reflector. Square wave modulated near-mm wave radiation causes temperature and pressure modulation in the gas cell to be picked up by a microphone and the effect of absorbed dissipation is nullified using ohmic dissipation. Very good agreement has been found [16] between the photo-acoustic power meter and the currently available UK power standard for waveguides (thermistor mounts characterised using waveguide microcalorimeters).

2.4 Light scattering and spectroscopic techniques

Light scattering is a term referring to the physical processes involving the interaction of light with matter. Evaluation of the scattered light with regard to its intensity or wavelength often yields information about the scattering matter [17]. A simple technique that may be used to distinguish between inelastic and elastic scattering is observation of a change in frequency, $\Delta\omega/\omega_0$ between the incident and scattered light. Because of the proportional relation between energy, wavelength, wave-number and frequency, and the relation between energy and frequency through Planck's law, energy is often expressed in cm^{-1} or Hz units. Rayleigh scattering occurs in volumes of randomly oriented particles and is used extensively in heat and mass transfer measurements. Rayleigh scattering deals with particles which are small compared to the wavelength in such a way that their interaction does not appreciably change the phase or amplitude of the light wave, and the scattered light intensity is proportional to the square of the particle diameter and inversely proportional to the fourth power of the wavelength of light (for a comprehensive treatment, see [18]).

Bragg scattering on the other hand is observed in structured arrangements of particles such as crystals, provided that the distance between adjacent particles is of the same order as that of the incident wavelength. The strongest intensity is observed if the Bragg condition is fulfilled, i.e. the light is incident at the Bragg angle to the scattering plane. Because the characteristic distance between the atoms in a lattice structure is of the order of few Ångströms, electromagnetic waves from the X-ray region are used, and such a method is commonly known as X-ray diffraction. A summary of different measuring techniques based on light scattering processes is shown in Table 2.1.

In order to explain the Rayleigh scattering process, it can be initially assumed that the molecules of a substance are substantially motionless and ordered, in a strict lattice structure, and therefore an incident mono-chromatic light beam of frequency ω_0 will produce no resultant scattering intensity due to destructive interference, allowing the light beam to be visible only in the forward direction. For a random motionless lattice structure, however, there will be some resultant scattering intensity of constant value. This intensity is less than that predicted by Rayleigh theory since there is still a great amount of destructive interference. The frequency spectrum of the scattered light reveals a sharp line at the incident light frequency ω_0 provided that Rayleigh scattering is the only scattering process involved.

For most applications, however, the thermal or Brownian random motion of molecules is responsible for the temporal fluctuations of the scattered light intensity and the broadening of the spectrum of the scattered light [19, 20]. The temporal intensity fluctuations observed are generally due to density fluctuations in the medium around the macroscopic equilibrium density. Such observations, therefore, yield information on transport properties such as the thermal diffusivity, diffusion coefficient, sound absorption coefficient and speed of sound in the medium. Digital correlation or a spectrum analyser can be used to analyse these density fluctuations [21]. When a non-periodic signal is correlated over a long period T, the correlation function is a first order decaying function. Usually the correlation function is required in its normalised form and the

Table 2.1   A summary of different measurement techniques based on light scattering and typical applications

Measurement technique	Scattering process	Typical application
Laser Doppler velocimetry	Mie	Particle size, particle velocity
Raman spectroscopy	Raman	Molecular concentration, temperature
X-ray diffraction	Bragg	Density
Rayleigh thermometry	Rayleigh	Temperature, density
Photon-correlation spectroscopy	Rayleigh	Thermal conductivity, diffusion coefficient, molecular structure
Interferometry	Brillouin	Sound velocity, sound absorption
Fluorescence spectroscopy	Fluorescence	Density of atoms and molecules
Absorption spectroscopy	All processes	Concentration of atoms and molecules

value of the characteristic decay time τ_c of this exponential function contains information on the parameters under investigation.

Practical digital correlators have a large number of sampling channels which represent distinct points of the correlation function, each successive channel representing an increasing time lag τ in the function. The actual correlation is achieved by means of shift registers (one for each channel). Alternatively, information on the density fluctuations can be extracted by Fourier transformation of the temporal intensity fluctuations into their frequency components, thus obtaining the power spectral density of the signal. This transformation is known as the Wiener–Khintchine theorem, and results in a bell-shaped Lorentzian distribution curve centred at ω_0 with a line width Γ proportional to the density fluctuations. Since the power spectrum and the corresponding correlation function are Fourier transform pairs, Γ is directly related to the characteristic decay time $\Gamma = 1/\tau_c$ [22]. In practice, spectrum analysers are used to obtain the integrated power density of the signal at a set frequency ω. By tuning the filter through a frequency range, the power spectral density may be measured.

Very often, however, the Rayleigh contribution (which represents local entropy fluctuations at constant pressure) to the scattered light spectrum is accompanied by a doublet known as the Mandelshtam–Brillouin doublet, due to adiabatic pressure fluctuations which give rise to sound waves travelling through the fluid. Light scattered from these waves is slightly frequency shifted in both directions in analogy to the Doppler effect, the amount of frequency shift being dependent on the local speed of sound and the scattering vector. When scattered light intensities are high a homodyne method is generally employed, whereas when low fluctuation intensities are to be measured signal enhancement is achieved by superimposing a second coherent beam or local oscillator of constant intensity with the scattered light beam.

Alternatively to Rayleigh light scattering measurements, Raman spectroscopy [23] may be applied when information on a molecule must be extracted with regard to its possible energetic (vibrational and rotational) states. For every molecule, there are discrete rotational and vibrational energy levels which cause a frequency shift between the scattered and incident light, forming therefore a unique signature for the molecule at a particular energy level.

Upon absorption of one photon of the incident monochromatic radiation a molecule is lifted to a virtual state above the stable states of the ground electronic state. Within a very short period of time (10^{-14} s) the molecule returns from this state to a stable energetic state by emitting one photon of light. If the final molecular state is above the original state, the wave-number of the emitted photon shifts to lower values (higher wavelength, red shift), and if the molecule is originally in an elevated energetic state and returns to a lower state, a blue shift occurs. In spectroscopy, a red shift is called a Stokes transition and a blue shift is called an anti-Stokes transition.

The intensity of an observed Raman signal is proportional to the differential cross-section of the molecule, the solid angle of observation, the observed length of the measuring volume, the number density of the molecules in the initial energy level and the irradiance of the source. It therefore follows that temperature measurement is possible using Raman scattering as the population distribution of the energy levels of the molecules is described by the Boltzmann curve. Such information can be obtained by either rotational or vibrational Raman spectroscopy.

Furthermore, with Raman scattering it is also possible to perform simultaneous investigations of different species in a single measuring volume regardless of the physical state of the material. In heat and mass transfer problems, the media are usually either liquid or gaseous, while chemical and biological applications often deal with solid samples. The most widely used methods are spontaneous Raman scattering and coherent anti-Stokes Raman scattering, the latter providing higher intensity light signals useful for gaseous applications. Unfortunately the Raman signal signatures, although very precise on their information content regarding the energy state of a molecule, are very weak, requiring sophisticated photon converters such as photomultipliers, photodiode arrays or charge-coupled devices. An alternative to Raman measurement techniques based on laser induced fluorescence is one of the newest measurement techniques, and can provide the high signal strength often required [24–26].

The principal applications of laser induced fluorescence are measurements of minority species concentrations in complex reaction systems and temperatures, although pressure and velocity distribution measurements have also been described in the literature. The measurement is based on the natural fluorescence of atoms. Because this fluorescence occurs in transitions from weakly populated excited electronic energy levels, the signals are as weak as the population density. The effect of the laser induction process is to promote a considerable number of particles, molecules or atoms from the densely populated lower energy levels to the excited levels. Therefore, the subsequent emission of radiation by fluorescence is strongly enhanced. Besides the standard linear laser induced fluorescence, there are laser induced saturated fluorescence and laser induced pre-dissociation fluorescence. These techniques provide better measurement accuracy, yielding higher evaluation and interpretation of the scattered intensities. Due to the large signal intensities and the high achievable spatial resolution the main field of application of laser induced fluorescence is the two dimensional imaging of reacting flows. However, a major limitation of this type of spectroscopy is its restricted application to ground states only, and if excited states must also be investigated, absorption spectroscopy would be preferable.

A major advantage of absorption spectroscopy is that it utilises all scattering processes and is therefore capable of providing more information about the state of a sample. The general method for measuring absorption spectra is based on the determination of the absorption coefficient from the spectral intensity which is transmitted through an absorbing path. In most cases when absorption spectroscopy is performed, the subject of interest is either the absolute concentration of one or more chemical species present in a mixture or the temperature of the mixture. In the first case, a spectral window with dominant absorption must be found whereas in the second, the absorption line must also have a maximised temperature dependence. The macroscopic description of the absorption process is the Bouguer–Lambert–Beer law:

$$I(\omega, z) = I_0 e^{-a(\omega)z} \qquad (9)$$

where $a(\omega)$ is the linear absorption coefficient, and z is the thickness of the absorbing medium. The linear absorption coefficient is related to the spectral absorption cross-section $\sigma(\omega)$ which is a measure of the area blocked by each molecular absorber based on geometrical optics $\sigma(\omega) = a(\omega)/n$, where n is the number of absorbers per unit volume. However, care must be taken in the application of the above formula, as for high incident power levels, the absorption is non-linear due to a change in population of absorbers in the ground state. As the lifetimes of the excited states increase dramatically with increasing wavelength, care must be taken to avoid non-linearities, especially towards the mid- and far- infra-red regions of the spectrum.

The parameters of interest that characterise absorption line spectra are the position, the shape, the width and the strength. Position and strength are intrinsic properties of the specific molecule under investigation in a mixture, whereas width and shape are very strongly influenced by the environment and the overall chemical composition of the mixture. Although theoretical calculations of molecular line positions are often possible, absolute line positions and strengths are usually obtained from the literature as well as some spectral line data banks.

Increasing temperature and pressure tend to increase linewidth and limit the spectroscopic resolution of the measurement. For an isolated absorption line of centre frequency ω_0 the linear absorption coefficient $a(\omega)$ can be split into the product of a frequency independent line strength factor (in units of cm^{-2}) and a line shape function which performs a normalisation procedure (and therefore has units of cm).

The introduction of a line strength parameter offers the possibility of treating these broadening effects separately as a line shape function and as a half width at half maximum (HWHM) function. The HWHM function depends on the amount of energy to which the molecules are subjected. Using Heisenberg's uncertainty principle, $\Delta E \Delta t \simeq h$, it follows that there is a large uncertainty of the energy level of a molecule when it is in an excited state of finite lifetime. This results in a natural linewidth broadening, which increases dramatically from the far-infra-red to the ultraviolet and has a Lorentzian shape. In addition, the thermal motion of the molecules contributes to a Doppler broadening effect that forces the lineshape to have a Gaussian profile. Finally, perturbations in the surrounding medium during the interaction between photons and the absorber give rise to a collisional broadening effect. However, experimental lineshape functions are better described as the convolution between a Gauss-weighted sum with shifted Lorentzian profiles, the so called Voigt function.

The minimum detectable concentration of absorbing molecules is line strength dependent and is determined by the absorption path length z, the absorption cross-section $\sigma(\omega)$ and the minimum detectable relative intensity change $\Delta I/I_0$ caused by the absorption. It follows that, in order to reach a high detection sensitivity for absorbing molecules, $z\sigma(\omega)$ should be large and the minimum detectable value of $\Delta I/I_0$ as small as possible. When the absorption coefficient is small, measurements cannot be very accurate since a small difference between two large quantities must be achieved in the presence of noise. However, there are several techniques used in absorption spectroscopy that adopt some sort of modulation technique for noise reduction.

Provided that the absorbers have a significant magnetic or electric dipole moment, instead of tuning the laser, the absorption line itself can be continuously tuned via an external magnetic field (Zeemann effect) and this is known as laser magnetic resonance [27]. If an external electric field is used instead, the technique is called Stark spectroscopy [28].

Further to the previously described noise reduction techniques, when lasers with discrete but very densely laying lines such as CO_2, HF and CO are used, a differential absorption technique can be utilised. In this case, a pair of interference free laser lines have to be found in such a way that only one line is on resonance and the other is off resonance, so that the concentration of the absorbers can be measured by comparing the differential absorption, on and off the absorption line.

More recently, there has been a lot of interest in indirect absorption detection techniques where the signal is not derived from photon detection but is transferred to a quantity which can be measured with higher accuracy or better resolution. A typical example is photo-acoustic spectroscopy, whereby the absorption of photons and subsequent relaxation of the excited molecules induces a temperature change in a gaseous medium and hence a pressure change in a closed absorption cell that can be detected with a sensitive microphone. Alternatively, the long lifetimes of rovibrationally (rotationally and vibrationally) excited molecules or atoms in very low pressure media can be detected using optothermal spectroscopy, whereby the energy stored in the absorber is detected using a bolometer.

Another powerful method, particularly useful in analysing the chemistry of gas discharges and flames, is optogalvanic spectroscopy [29], where the change in the discharge current is monitored while tuning the wavelength of light sent through plasma. This method is especially valuable for investigating unstable, highly reactive intermediate reaction species. The selectivity of this method can still be improved using velocity modulation spectroscopy, whereby a periodic modulation of the discharge current is applied. The periodic Doppler shift caused by this velocity modulation can be extracted with a lock-in amplifier and can be used to distinguish between positively and negatively charged particles (opposite phase of the lock-in signal) while signals of neutral particles show no Doppler shift and hence no periodicity at all. Perhaps the highest sensitivity possible (down to single particle detection) can be achieved via ionisation spectroscopy [30], in which the detected ions are produced by subsequent ionisation of the excited absorber.

Alternatively, the absorption process may take place within a laser cavity or into a cavity strongly coupled to it, in which case the absorbance can be monitored by recording the laser output power in conjunction with the variable output wavelength. Such intracavity spectroscopy can be further combined with the previously described methods, enhancing their specific advantages with the very high sensitivity of the intracavity set-up. As stated earlier, if an intracavity set-up is not desirable, optical isolation of the laser source and Brewster angle windows in the absorption cells are necessary in order to ensure spectral purity of the source and minimal Fabry–Perot interference from the expansion of the cell during operation. Finally, in order to circumvent the lower resolution limit imposed by Doppler broadening, particularly for the visible and UV transitions, powerful techniques such as saturation, polarisation and multi-photon spectroscopies are currently under development as a new class of non-linear absorption spectroscopies [31] in many laboratories around the world.

Further to amplitude modulation, noise reduction by harmonic detection can be used if either the laser or the absorption line can be tuned periodically in the kHz range. This is most commonly achieved by superimposing the modulation on the injection current of the laser diode. If the modulation amplitude is sufficiently small, the transmission function can be expanded as a Taylor series around the central wave-number of the emitted light, so that frequency independent background noise, cell window noise, laser intensity noise or sample density fluctuation noise can be minimised using phase-sensitive detection. This technique is mostly known as derivative spectroscopy and allows the modulation used in the harmonic detection scheme to shift the absorption signal from dc to higher harmonics of the modulation frequency. Since the net gain for signals not matching the modulation frequency or higher harmonics thereof is much smaller, this will result in a net noise reduction. Further increase in the modulation frequencies in the hundred MHz region reduces $1/f$ noise and is theoretically capable of quantum limited detection, although demodulation can only be performed using mixers (down-converters) and other HF-components.

A relatively new method of high frequency modulation spectroscopy using low frequency detection is two-tone frequency-modulation spectroscopy, whereby the laser output is phase modulated in the GHz range using an electro-optic crystal, while it is amplitude modulated at frequencies in the MHz range. The detector output is then fed into a frequency mixer and the final signal is received using a lock-in amplifier in the kHz range.

If wavelength tuning is possible, an alternative to the harmonic technique is provided by sweep integration. Using successive scans of absorption spectra, the statistical noise amplitude (rms) in any wavelength channel increases only with the square root of the scan number whereas the signal increases linearly, leading to an overall improvement of the signal-to-noise ratio. Such a technique also offers the possibility of resolving time dependent concentration changes by operating on a single shot basis. In experiments using pulsed lasers, with pulse durations the order of nanoseconds or less, the time gated noise reduction technique, in conjunction with co-averaging over several pulses, produces a similar noise reduction efficiency to the sweep integration technique.

Alternatively, the time gated signal-averaging procedure can be used to perform integrative spectroscopy. In this case, the laser is repeatedly scanned over a single isolated absorption line and the recorded absorption signal is time integrated. By ensuring a linear frequency scan, the time integration of the signal corresponds to a frequency integration of the absorption. This value is proportional to the signal strength, or alternatively, provided that this quantity is known, the result is a direct measurement of the absolute concentration provided that most of the absorption line can be covered.

2.5 Advances in optical properties and polarisabilities of atoms and molecules

In the previous section, attention was given to the interaction of light with matter. Of special interest to these interactions is the charge redistribution that occurs when a particle is exposed to an electric field. In the presence of such a field, a new charge distribution occurs in atoms, molecules and clusters. It is common to refer to the lowest-order dipole moment of a neutral particle as its dipole moment. Provided that a uniform electric field is applied to a particle, its dipole moment becomes

$$p = p_0 + \alpha_p E + \dots \tag{10}$$

where p_0 is the permanent dipole moment, α_p is the lowest-order induced dipole moment in the species (polarisability) and E is the electric field strength. A particle's polarisability has units of volume and is of the same order of magnitude as its volume (e.g. 10^{-23} cm³ for a single atom). This is of interest to instrumentation because it is an indicator of physical size, structure and shape, it is involved in collision phenomena between neutral particles and other neutral or charged particles, and it determines the response of neutral particles to fields produced by lasers.

Since particles can absorb as well as emit photons, the polarisability consists of two parts: a dispersive part which accounts for absorption and instantaneous re-emission of photons which corresponds to scattering of light, and an absorptive part which accounts for absorption of light by the particle with subsequent de-excitation due to spontaneous emission or collisional quenching. Furthermore, there is a frequency dependency of the light that induces the dipole moment. In high frequencies, the charges are unable to follow the changing field and the polarisability drops to zero.

The dispersive and absorptive parts of the polarisability are related to each other through the Kramer–Kronig relation. This relation has its origin in linear systems theory and therefore only accounts for a linear response of matter to an applied time-varying electric field. As such, it only applies to Rayleigh and not to Raman transitions. Since the polarisability is the response of a particle to an external field, it follows that the polarisability also induces a change in the energy of the field, i.e. a Stark shift.

2.5.1 Standard techniques in the measurement of polarisabilities

Traditional techniques for measurement of average polarisabilities consist of measuring the dielectric constant ε of the material. This can be conveniently done by comparing the material capacitance with a known capacitance using an ac bridge circuit where the differential signal is fed through a charge amplifier and a phase sensitive detector. The polarisability is then related to the Debye equation

$$\alpha_p = \frac{3}{4\pi N}\left(\frac{\varepsilon-1}{\varepsilon+2}\right) - \frac{\mu_0^2}{3kT} \tag{11}$$

where N is the number density and μ_0 is the permanent electric dipole moment of the molecule. In the special case where $\mu_0 = 0$ the expression is called the Clausius–Mossotti relation. Measurements of the dielectric constant can provide estimates of static or near-static polarisabilities of gases or solids.

Alternatively, the ratio of the speed of light in a vacuum to the speed of light in the material (which is the refractive index n of the material) may be used to deduce the polarisability from the Lorentz–Lorenz equation:

$$\alpha_p = \frac{3}{4\pi N}\left(\frac{n^2-1}{n^2+2}\right) \tag{12}$$

This equation is valid for non-polar molecules or at sufficiently high frequencies where the permanent dipole moment cannot follow the varying electric field. A dipole moment term is added when the frequency of the radiation is less than or comparable to the rotational frequencies of the molecule. The Lorentz–Lorenz and Clausius–Mossotti equations are related by the Maxwell relation $\varepsilon = n^2$. Measuring the index of refraction using Snell's law and observing the angle of refraction at the interface between a sample and a reference material determines the polarisability at the frequency of the light used to make the measurement. The dc polarisability is usually determined using long-wavelength radiation or by extrapolating short-wavelength measurements to dc.

Alternatively to the refractive index technique, it is possible to calculate the polarisabilities of gas molecules from Rayleigh scattering experiments. Generally, the oscillating electric field of the incident light induces an oscillating dipole moment in the molecule, which, as a result, re-radiates light in a dipole pattern. In spherically symmetric molecules, the induced dipole moment is parallel to the incident polarisation and is proportional to the polarisability of the particle. For non-spherically symmetric molecules, however, the induced dipole moment depends on the orientation of the molecule and therefore Rayleigh scattering causes depolarisation of the incident light. The depolarisation function defined as the ratio of the maximum to minimum steradians in the direction of scattering is the most commonly used parameter in such studies.

In liquid samples, the electro-optic Kerr effect may also be used for the measurement of polarisabilities. A strong orienting electric field is used to induce optical birefringence in the sample so that different responses may be obtained for light propagating perpendicularly or parallel to the orienting field.

Alternatively, it is possible to determine the polarisability of a molecule (especially of alkali metals) by observing the deflection of a molecular beam passing through a static inhomogeneous electric field. The spatial deflection of the beam observed on a hot wire detector is inversely proportional to the square of the velocity of the atom or molecule. The technique uses an oven to evaporate the sample and a rotating disk velocity selector to reduce the ambiguity in the initial velocity of the molecules.

A variance to this technique is the *E–H* gradient balance technique in which the electric force caused by the interaction of the induced dipole moment with the electric field gradient, is balanced by a magnetic force caused by the interaction of an effective magnetic dipole moment with a magnetic field gradient. An advantage of this technique is that the observed signal is independent of the velocity distribution in the beam. It requires, however, that the atoms or molecules have an appropriate ratio of polarisability to magnetic dipole moment for the balancing of the forces to occur.

A further variant of the technique which permits very precise measurement of the polarisability to be performed is called atomic beam resonance. In this technique, a magnet pre-selects molecules having a particular magnetic sub-level and allows them to interact with a magnetic and an electric field within a cavity. A second magnet then selects only the molecules that have undergone a particular transformation to reach the detector.

2.5.2 Recent advances in the measurement of polarisabilities

The most promising emerging new techniques for the determination of polarisabilities are dispersive Fourier transform spectrometry (DFTS) [32–34], the M-lines technique [35], position sensitive time of flight [36], the light force [37] and atom interferometry [38].

Dispersive Fourier transform spectrometry (Figure 2.9) is an interferometric technique that is used to measure the complex refractive index of gases, liquids and solids.

In DFTS, a broadband source is used and a mirror is driven in one direction through the zero path difference position of the interferometer so that the detector records an interferogram that is the squared modulus of the autocorrelation function of the time-dependent field of the input beam. Fourier transformation of the interferogram gives the power spectrum of the source. Placement of a sample in one of the partial beams modifies the interferogram. The ratio of the complex Fourier transforms with and without the sample in place is known as the

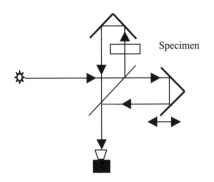

Figure 2.9   Single pass dispersive Fourier transform spectroscopy for the determination of optical constants

complex insertion loss of the sample, its modulus being the sample transmission spectrum. The variation with frequency of the complex optical constants of the sample can be determined from the complex insertion loss. This is further related to the complex transmission coefficient $\hat{t}(k)$ of the specimen,

$$\hat{L}(k) = \hat{t}(k)e^{-ikd} \tag{13}$$

(where k is the wave-number) the factor e^{-ikd} arising because the specimen replaces a length d of air in the arm of the interferometer. The complex refractive index of a sample is related to the measured complex insertion loss by

$$\hat{L}(k) = \frac{4\hat{n}}{(1 + \hat{n})^2} \frac{e^{ik(\hat{n}-1)d}}{1 - [(1 - \hat{n})/(1 + \hat{n})]^2 e^{2ik\hat{n}d}} \tag{14}$$

In experiments where a single pass transmission through the sample occurs, and provided that the sample does not act as a Fabry–Perot resonant cavity, closed form expressions for the complex refractive index can be found. The real and imaginary parts of the complex refractive index of the sample given by $\hat{n}(\omega) = n(\omega) + i\kappa(\omega)$ are related to the complex insertion loss function from

$$n(k) = 1 + \frac{1}{kd}[\arg(\hat{L}(k)) \pm 2m\pi] \tag{15}$$

and

$$\kappa(k) = \frac{1}{kd}\ln\left[\frac{4n(k)}{[1 + n(k)]^2} \frac{1}{|\hat{L}(k)|}\right] \tag{16}$$

where $m = 1, 2$, etc. The complex refractive index is related to the mean polarisability through the Lorentz–Lorenz equation and avoids the Kramer–Kronig relation, which calculates the imaginary part of the polarisability from knowledge of the real part of the polarisability over a large wavelength range. The complex refractive index is directly related to the complex polarisability from

$$\hat{a}_p = \frac{3}{4\pi N} \frac{(\hat{n}^2 - 1)}{(\hat{n}^2 + 2)} \tag{17}$$

Similarly to the complex refractive index, the complex polarisability has a real and an imaginary part given from $\hat{a}_p = a'_p + ia''_p$. It follows that

$$a'_p = \frac{3[(n^2 + \kappa^2)^2 + n^2 - \kappa^2 - 2]}{4\pi N[(n^2 + \kappa^2)^2 + 4(n^2 - \kappa^2 + 1)]} \tag{18}$$

and

$$a''_p = \frac{9n\kappa}{2\pi N[(n^2 + \kappa^2)^2 + 4(n^2 - \kappa^2 + 1)]} \tag{19}$$

An alternative to the DFTS is the M-lines technique. This is generally used for materials whose refractive index is less than that of the substrate that supports them and has allowed researchers to measure the polaris-ability of large molecules formed as thin films. It relies on the observation of several lines that appear simultaneously when light is optimally coupled from a prism into a thin film (Figure 2.10).

The position-sensitive time-of-flight technique is useful for measuring dc polarisabilities and is based on the traditional approach of deflecting a neutral beam using a non-uniform electric field. It uses, however, a sensitive time-of-flight spectrometer and laser photo-ionisation. Clusters of atoms from a beam are initially deflected by an amount proportional to their polarisabilities and inversely proportional to their mass. Once spatially separated, they are ionised by absorbing energy from an excimer laser and are further accelerated by an electric field in a direction orth-ogonal to the original direction of propagation. This procedure further spatially separates the ion clusters. An ion detector in conjunction with a gating technique is then used to isolate individual clusters of ions.

For ac polarisability measurements, the light force technique may be used. This relies on the observation of a change in velocity distribution of an atomic beam. A standing wave laser perpendicular to the direction of propagation of the atomic beam is used to accelerate the atoms. The change in the velocity distribution of the atoms is measured by probing the beam with a laser at a frequency near resonance, Doppler shifted atoms showing a significant difference in absorbance.

Finally, the newly emerging field of atom interferometry has shown potential for measuring dc polarisabilities of atoms. The technique has been demonstrated using a sodium beam and a grating interferometer, where by applying an electric field in one of the separated beams, a Stark shift due to a change in the potential energy of the quantum-mechanical wave-function is observed. The phase difference between the two atomic beams is related to the potential energy corresponding to the Stark shift

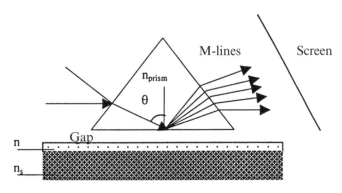

Figure 2.10   Varying the incidence angle to the prism, light is coupled by the prism into a lossless guided mode in the thin film

which, in turn, is directly proportional to the polarisability of the atom and the square of the electric field applied.

Although the polarisability is a fundamental electronic property of a single particle and determines common optical properties such as the refractive index and absorption coefficient of a system of particles, it is also possible to use its macroscopic equivalent which is known as the electric susceptibility χ. The susceptibility describes the scattering process assuming a collection of particles. For an isotropic electronic response of a system the complex refractive index $\hat{n}^2(\omega) = n(\omega) + i\kappa(\omega)$ of the material is directly related to the complex susceptibility by

$$\hat{n}^2(\omega) = 1 + 4\pi\hat{\chi}(\omega) \tag{20}$$

where $\hat{\chi}(\omega) = \chi'(\omega) + \chi''(\omega)$, which implies that $1 + \chi'(\omega) = n^2(\omega) - \kappa^2(\omega)$ and $\chi''(\omega) = 2n\kappa(\omega)$. Choosing the frequency of the light to be close to that of a resonance for the particles, maximum absorption occurs so $n = 1$ and $\chi' = 0$. For frequencies away from resonance, $\kappa \cong 0$ and $\chi'' \cong 0$ so that $\chi' = n^2 - 1$. Furthermore, the dielectric constant is a microscopic property which is also related to the susceptibility and the magnetic permeability μ of the medium by $\mu\varepsilon = 1 + 4\pi\chi$. The recent demand for memory devices as well as advances in the calculation of the non-linear susceptibility in ferroelectrics [39] highlights the importance of developing new techniques for accurately performing measurements of the non-linear optical coefficients.

2.6 Recent advances in instrumentation based on the interaction of light with matter

The use of the mechanical properties of light to manipulate matter has advanced considerably in the last decade. The use of optical tweezers to manipulate small biological structures, the use of light to trap and cool atoms as well as the use of atomic beams to create nanostructures and the atomic force microscope are among the most important recent advances [40–50].

It is now well known that atoms in an atomic beam will absorb photons from a counter-propagating beam and this will result in some photons being scattered. Although the photons are absorbed from one direction, the scattered photons are emitted in all directions. Hence momentum is absorbed by the atoms along the axes of the light. The scattering force is a dissipative force. The cooling laser is fixed in frequency and the atomic levels are Zeeman shifted so that the atomic resonance frequency is tuned as the atoms are moved through a gradient magnetic field. The motion of the atoms through a magnetic field gradient allows the tuning of the resonant frequency to compensate for the Doppler shift in resonant frequency caused by the cooling and slowing of atoms.

Alternatively, a laser can be continuously tuned to be resonant with the atoms moving towards it. In the Doppler cooling case, the force is proportional to the polarisability. These advances have assisted in the initial cooling for the formation of atomic Bose–Einstein condensates which requires, however, further evaporative cooling with an rf magnetic field to pump the hotter atoms in the Boltzmann tail from trapped states to unstable states. However, it is impossible to trap a small dielectric particle at a point of stable equilibrium in free space using only the scattering force of radiation pressure alone, and therefore a magnetic field gradient is used.

A magneto-optic trap (MOT) consists of the intersection of six laser counter-propagating beams in a configuration which allows the formation of three standing wave axes. The intersection produces a potential well in space that traps atoms in a small volume at the intersection. The force of an atom in an MOT consists of a combination of a restoring force due to the magnetic field gradient and a damping force corresponding to Doppler and molecular orientation cooling.

Although atom traps use the scattering force optically to confine atoms to a small region of space, optical traps (commonly called optical tweezers) rely on the gradient force to confine particles optically in a small region of space. Optical tweezers have recently been used in biology to move and manipulate small objects such as single cells, organelles and bacteria. An optical tweezer consists of a focused beam of laser light that traps objects just beyond the focal plane of the light. As particles seek regions in space where the light intensity is highest, only the gradient force contributes to lateral confinement, whereas an axial force on the particle is a combination of the gradient and scattering forces. At the position just below the focus of the beam, the gradient force and the scattering force point in opposite directions. The technique uses laser powers of the order of 1–100 mW and wavelengths from 0.5 to 2.0 mm. The resulting forces are in the range of 1–40 pN and the trapped object size varies between 0.05 and 200 mm.

Another area of current interest is that of nanostructure fabrication using atom-optics. The technique uses laser-focused atom deposition to create two dimensional arrays of atomic 'dots'. Initially an optical standing wave with the desired pattern is produced by two counter-propagating laser beams just above the substrate. The focused atoms traversing the standing wave are then deposited at the nodes of the field. The technique allows very regular arrays to be produced with accurate periodicity due to the optical standing wave used to form the pattern. Atom-optics techniques have found application in the production of arrays as a calibration standard for lithographic processes as well as in the production of very regular arrays of small magnetic elements.

Finally, another recent instrument which is now used to observe interactions in matter is the atomic force microscope. This is capable of creating three dimensional surface images with sub-nanometre resolution and operates by measuring the deflection of a sharp probe tip as it approaches and retracts from the surface under observation. Its operation is defined by long and short range tip surface forces. The local force experienced by the tip is translated into a bending of a cantilever which can be measured by monitoring the angle of reflection of a laser beam off the cantilever end. Feedback variances of this probe are currently under development in many laboratories around the world.

2.7 Recent advances in THz measurement and instrumentation systems

Another interesting emerging technology is that of imaging systems using ultrafast pulses in the far-infra-red part of the electromagnetic spectrum. Such pulses are generally produced using either photoconduction or optical rectification. In photoconduction, charge carriers are created using the light beam from a femtosecond laser focused on a very tight spot between two electrodes 100 mm apart which are laid on a silicon or GaAs substrate. These are accelerated using an electric field so that a transient photocurrent that radiates electromagnetic waves is produced. An antenna structure ensures the coupling of the radiative modes from the structure into free space. The conversion efficiency of visible pulse energy to THz radiation is only of the order of 0.1% and therefore THz pulses of the order of only 1 nJ may be produced using high repetition rate (250 kHz) femtosecond lasers. Using low repetition ultrafast laser systems (10–1000 Hz), energies of the order of 1 μJ have been reported.

The difference between optical rectification and photoconduction is that the visible exciting beam creates virtual rather than real carriers. Because in photoconduction the second order susceptibility of the crystal is used for difference frequency mixing, the electric field of the THz pulse has the same shape as the intensity envelope of the visible exciting pulse. For optimum conversion from visible to far-infra-red wavelengths, one has to match the group velocity of the visible pulse with the phase velocity of the THz pulse.

The reverse of optical rectification is electro-optic sampling [51, 52]. A THz pulse incident on an electro-optic crystal such as ZnTe induces birefringence through the Pockels effect. An ultrafast visible probe pulse with a variable delay co-propagating through the same crystal experiences a retardation that can be observed with a balanced detection scheme. By scanning the relative time-delay of the probe pulse, a time-domain record of the THz pulse may be produced [53]. A refinement to this method that has been recently reported is to use a chirped probe pulse to sample the electro-optic sampling process so as to observe the relation between wavelength and relative time delay [54]. Detecting this probe pulse spectrometrically with a diode array detector allows one to measure the entire THz pulse shape in a single laser shot (Figure 2.11).

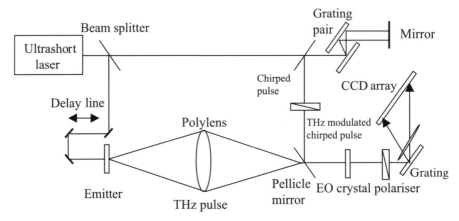

Figure 2.11   Set-up for THz generation and electro-optic sampling of a terahertz pulse with a chirped probe beam, as described in [54]

Although bandwidths for a THz pulse of 37 THz have been reported, the time bandwidth relation $\Delta\omega\Delta\tau = 0.32$ suggests that with the shortest visible pulses achievable (4–5 fs) power at frequencies from 0 to 160 THz is possible and a source covering the entire far-infra-red spectrum up to $\lambda = 1.8\mu m$ is possible. THz spectroscopy is currently offering some exciting possibilities for studying the absorption spectrum of water and other liquids of chemical and biological importance.

As mentioned earlier, an important aspect of femtosecond THz pulses is their short duration and a possible synchronisation with other visible light sources. This makes it possible to perform time-resolved pump-probe experiments for studying chemical reaction dynamics or the conduction and trapping properties of carriers in semiconductors, semiconductor superlattices and superconductors. More specifically, the excitation of a semiconductor above the band-gap with a visible ultrafast pulse creates a plasma of conduction band electrons. This gives rise to a transient absorption which when measured can give direct information about the electron mobility and the carrier lifetime. In addition, by pumping above the band-gap, information about the rates at which the carriers scatter between the different valleys can be obtained. The recent surge in the study of Bloch oscillations and the use of femtosecond visible excitation pulses to create coherent superpositions of Wannier–Stark exciton states in superlattices as well as the breakup of Cooper pairs and the collapse of the superconductive gap in superconductors can be probed using THz pulses. Finally, apart from time-resolved experiments, which use relatively low power THz pulses to observe reac-

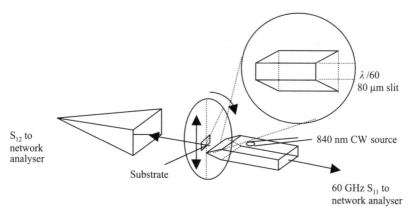

Figure 2.12 *Sub-wavelength scanning near field microscopy with a slit type probe as described in [55]. The use of optical fibres at the probe tip for the photo-excitation of free carriers in the substrate is also shown*

tions, the use of high power pulses in the future could offer the possibility to interfere with chemical reactions actively.

Another area of considerable interest in current research is that of imaging in the far-infra-red part of the spectrum. In conventional micro-scopy, diffraction effects limit the resolution of focal plane imaging to $\lambda/2$. Sub-wavelength resolution has been demonstrated using near field microscopes fitted with point-type probes. Recent advances in near field mm-wave microscopy, however, have shown that a metal slit $\lambda/60$ wide at the end of a waveguide operating above the cut-off frequency of the fundamental waveguide mode can provide stronger signals than those obtained using point-type probes. In order to achieve sub-wavelength resolution the sample is scanned using a combination of a rotational and translational movement with the slit as a scanning probe (Figure 2.12). Finally, an image reconstruction algorithm based on computerised tomographic imaging is used. Applications of this new technique include observation of photo-excited free carrier phenomena in silicon substrate as well as imaging defects and doping profiles. Advances in the response time of network analysers (0.4 ns) show promise for dynamic imaging of photo-excited free carrier phenomena in the future.

2.8 Optical and microwave waveguides

Of fundamental importance to the development of measurement sys-tems based on the interaction of light with matter is our ability to guide accurately as much energy as possible to the location where the transduc-tion phenomenon must occur as well as to reliably transmit this informa-tion to another convenient location. At relatively low frequencies, up to 1 GHz, wire circuits are capable of transmitting radio waves, whereas in the microwave part of the electromagnetic spectrum up to a frequency of 300 GHz, rectangular and microstrip waveguides are preferred. Rectangular waveguides have a restricted bandwidth of one octave within which they operate single-moded, and therefore systems operating in this frequency range have a restricted bandwidth. Furthermore, because at frequencies above 100 GHz the size of rectangular waveguides must be reduced and watchmaker precision is required to achieve the desired surface finish during the manufacturing process, components tend to be expensive and not widely available. Recent advances in photolithographic techniques, however, using masks and novel photoresist materials, have shown great promise in mass-producing cheap waveguides with superior transmission properties [56]. In the sub-millimetre part of the spectrum (300 GHz – 1 THz) there are no off-the-shelf rectangular waveguides available, and metal-insulator-semiconductor transmission lines on special substrates, e.g. cyclotene, are used. At still higher frequencies and up to 300 THz, where diffractive spreading of free-space propagating beams is not an issue, mirrors are the only means for guiding electromagnetic radiation, and systems are severely limited due to atmospheric absorption. In addition, lenses show large absorption above 1 THz and therefore are not suitable for guiding radiation. Above 300 THz, however, optical fibres may be used.

From this description, it follows that it is possible to treat uniformly the propagation of electromagnetic radiation in all waveguide types as special cases of the solutions of the Maxwell equations. It is common to use parameters such as the characteristic impedance to describe propagation inside a waveguide at microwave frequencies and refractive index to describe the propagation of electromagnetic radiation in the infra-red and visible part of the spectrum. In the far-infra-red part of the spectrum, it is useful to visualise the refractive index parameter as an impedance mismatch condition. A unified treatment of waveguides can be found in the book by Cronin [57].

Because of the wide availability of components and systems in the optical and infra-red parts of the electromagnetic spectrum, a very large proportion of measurement systems makes use of optical fibres. When light travelling through a medium with a refractive index n_1 encounters another medium having a smaller refractive index n_2 some of the light will be refracted, and some of it will be reflected, depending on the incidence angle. The angle of emergence of the refracted ray θ_2 can be found from Snell's law, $n_1\sin\theta_1 = n_2\sin\theta_2$, where θ_1 is the angle of incidence. If θ_1 is gradually increased, there will be a specific point where the angle of emergence is 90 degrees, at which point there is no partial reflection. This is known as the critical angle θ_c. If the angle of incidence is further increased, then eventually, all the light rays will be internally reflected. It follows that there is a discrete range of angles of incidence in an air/fibre interface that allow the principle of total internal reflection to occur, and this is called the acceptance angle θ_a. The value of $\sin\theta_a$ is known as the numerical aperture and is a basic parameter for fibre selection.

However, fibres are three dimensional objects, and as such, they support several modes of propagation, the fundamental two being meridional and skew. Skew rays follow light paths which never intersect the fibre axis: a special case is that of a ray which travels parallel to the fibre axis, never being reflected throughout the fibre length. Fundamental fibre theory is concerned only with meridional rays, which, as implied, travel through the fibre axis after rebounding many times in the core/cladding interface.

There is a wide range of monomode or multimode optical fibres available on the market, with a wide range of refractive index profiles (graded index, step index, parabolic) and different core-cladding dimensions. Although ray analysis is sufficient for explaining the performance of most of these fibres in amplitude modulating sensors applications, the more strict electromagnetic treatment of the subject is widely adopted by many researchers [58]. Furthermore, there has been increased interest using a variety of plastic optical fibres for sensor applications, mainly because they are cheap and easier to handle. Although transmission lengths have been limited to approximately 20 m in the past, recent advances in the graded index plastic optical fibres [59, 60], the development of high temperature plastic fibres [61] and perfluorinated fibres [62] are making them an interesting alternative to the well established glass fibre technology. Typically, the poly-methyl methyl acrylate fibres (PMMA), that are used in sensor applications, have a core-cladding diameter of 1.00 mm, a core refractive index (n_1) of 1.49, and a cladding refractive index (n_2) of 1.42, from which the values of numerical aperture (NA), acceptance angle θ_a, and critical angle θ_c can be calculated (Figure 2.13). The light transmission window for these fibres is at the visible region of the spectrum. For a PMMA fibre of 1 mm core diameter, the numerical aperture is given by $NA = \sqrt{n_1^2 - n_2^2} = 0.45$ and the critical angle and acceptance angle can be calculated from $\theta_c = \sin^{-1}(n_2/n_1) = 72.37$ and $\theta_a = \sin^{-1}(NA) = 26.83$, respectively.

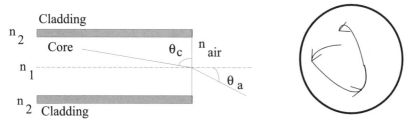

Figure 2.13   *Transmission of light at the air/fibre interface and ray projection showing the path of a general ray in the core of a graded-index fibre*

Optical fibres have been adopted in many sensing schemes because of their light weight, small size, immunity to electromagnetic interference, large bandwidth, immunity to vibration and shock, high sensitivity and high temperature performance as well as electrical and optical multiplexing capability. Furthermore, component costs are currently driven by the large commercial communication and optoelectronic markets. Although it is customary to classify optical fibre sensors as extrinsic or intrinsic depending on whether the modulation takes place outside or inside the fibre, in the next chapter an alternative classification is adopted so as to highlight the direct link between the modulation technique used and the physical principle of operation in each transduction scheme.

2.9 References

1 MARTIN, D. H., and BOWEN, J. W.: 'Long-wave optics', *IEEE Transactions on Microwave Theory and Techniques*, 1993, **41**, pp. 1676–1690
2 BOWEN, J. W.: 'A solid state noise source for millimetre wave spectrometry', *International Journal of Infrared and Millimetre Waves*, 1996, **17**, pp. 479–491
3 FAULKNER, E. A., and HARDING, D. W.: 'A high-performance phase-sensitive detector', *Journal of Scientific Instrumentation*, 1966, **43**, pp. 97–99
4 FAULKNER, E. A., and GRIMBLEBY, J. B.: 'High-speed linear gate', *Electronic Engineering*, 1967, **39**, pp. 565–567
5 GRIMBLEBY, J. B., and HARDING, D. W.: 'A new high-performance phase-sensitive detector', *Journal of Physics E, Scientific Instruments*, 1971, **4**, pp. 941–944
6 FAULKNER, E.: 'The principles of impendance optimization and noise matching', *Journal of Physics E, Scientific Instruments*, 1975, **8**, pp. 540–553
7 IRVINE, J. J., HADJILOUCAS, S., KEATING, D. A., and USHER, M. J.: 'Automatic optical fibre feedback potometer for transpiration studies', *Measurement Science and Technology*, 1996, **7**, pp. 1611–1618
8 ALLEN, D. W.: 'In search of the best block: An update', in 'Frequency standards and metrology' (Springer Verlag, 1989)
9 TOMLINSON, W. J., and FORK, R. L.: 'Frequency stabilisation of a gas laser', *Applied Optics*, 1969, **8**, pp. 121
10 HOUGH, J., HILS, D., RAYMAN, M. D., MA, L. S., HOLLBERG, L., and HALL, J. L.: 'Dye-laser frequency stabilisation using optical resonators', *Applied Physics B*, 1984, **33**, p. 179
11 IKEGAMI, T., SUDO, S., and SAKAI, Y.: 'Frequency stabilisation of semiconductor laser diodes' (Artech House, 1995)
12 ZOZ, J., and BARABAS, U.: 'Linewidth enhancement in laser diodes caused by temperature fluctuations', *IEE Proceedings – Optoelectronics*, 1994, **141**, pp. 191–194
13 OLKKONEN, H.: 'Dual diode preamplifier for detection of laser Doppler interference photons', *Review of Scientific Instruments*, 1991, **62**, pp. 238–239

14 EPPELDAUER, G., and HARDIS, J. E.: 'Fourteen-decade photocurrent measurements with large-area silicon photodiodes at room temperature', *Applied Optics*, 1991, **30**, pp. 3091–3099

15 FONCK, R. J., ASHLEY, R., and DURST, R.: 'Low-noise photodiode detector for optical fluctuation diagnostics', *Review of Scientific Instruments*, 1992, **63**, pp. 4924–4926

16 MOSS, D. G., BIRCH, J. R., ADAMSON, D. B., LUNT, B., HOTGETTS, T., and WALLACE, A.: 'Comparison between free space and in-waveguide power measurement standards at 94 GHz', *Electronics Letters*, 1991, **27**, pp. 1134–1137

17 BERNE, B. J., and PECORA, R.: 'Dynamic light scattering' (John Wiley, New York, 1976)

18 VAN DE HULST, H. C.: 'Light scattering by small particles' (Dover Publications, 1981)

19 VAN VLECK, J. H., and WEISSKOPF, V. F.: 'On the shape of collision-broadened lines', *Review of Modern Physics*, 1945, **17**, pp. 227–235

20 HERBERT, F.: 'Spectral line profile: A generalised Voigt function including collisional narrowing', *Journal of Quantum Spectroscopy and Radiative Transfer*, 1974, **14**, pp. 943–951

21 PECORA, R.: 'Dynamic light scattering', in 'Applications of photon correlation spectroscopy' (Plenum Press, New York, 1985)

22 COVA, S., and LONGINI, A.: 'An introduction to signals, noise, and measurements in analytical laser spectroscopy' (Wiley Interscience, New York, 1979), pp. 412–488

23 LONG, D. A.: 'Raman spectroscopy' (McGraw-Hill, London, 1977)

24 EVERALL, N., JACKSON, R. W., HOWARD, J., and HUTCHINSON, K.: 'Fluorescence rejection in Raman spectroscopy using a gated intensified diode array detector', *Journal of Raman Spectroscopy*, 1986, **17**, pp. 415–423

25 ZHANG, Z. Y., GRATTAN, K. T. V., PALMER, A. W., and MEGGITT, B. T.: 'Pronys method for exponential lifetime estimations in fluorescence-based thermometers', *Review of Scientific Instruments*, 1996, **67**, pp. 2590–2594

26 SUN, T., ZHANG, Z. Y., GRATTAN, K. T. V., and PALMER, A. W.: 'Application of singular value decomposition in average temperature measurement using fluorescence decay techniques', *Review of Scientific Instruments*, 1998, **69**, pp. 1716–1723

27 ZINK, L. R., JENNINGS, D. A., EVENSON, K. M., SASSO, A., and INGUSCIO, M.: 'Stark spectroscopy using far infra-red radiation', *Journal of Optical Society of America B – Optical Physics*, 1987, **4**, pp. 1173–1176

28 WEBER, W. H., TANAKA, H., and KANAKA, T.: 'Stark and Zeeman techniques for laser spectroscopy', *Journal of Optical Society of America B – Optical Physics*, 1987, **4**, pp. 1141–1226

29 TRAVIS, J.C.: 'Analytical optogalvanic spectroscopy in flames', in MARTELLUCCI, S., and CHESTER, A. N. (Eds): 'Analytical laser spectroscopy' (Plenum Press, New York, 1985)

30 HURST, G. S., and PAYNE, M. G.: 'Principles and applications of resonance ionisation spectroscopy' (Hilger Publications, Bristol, 1988)

31 LETOKHOV, V. S., and CHEBOTAYEV, V. P.: 'Nonlinear laser spectroscopy' (Springer Series in Optical Science, 1997, vol. 4)

32 HOHM, U., and KERL, K.: 'A Michelson twin interferometer for precise measurements of the refractive-index of gases between 100 K and 1300 K', *Measurement Science and Technology*, 1990, **1**, pp. 329–336

33 BIRCH, J. R., and CLARKE, R. N.: 'Dielectric and optical measurements from 30 to 1000 GHz', *Radio and Electronic Engineer*,1982, **52**, pp. 565–584

34 PARKER, T. J.: 'Dispersive Fourier transform spectroscopy', *Contemporary Physics*, 1990, **31**, pp. 335–353

35 DING, T. N., and GARMIRE, E.: 'Measuring refractive index and thickness of thin films – A new technique', *Applied Optics*, 1983, **22**, pp. 3171–3181

36 HEER, W. A., and MILANI, P.: 'Large ion volume time-of-flight mass-spectrometer with position-sensitive and velocity-sensitive detection capabilities for cluster beams', *Review of Scientific Instruments*, 1991, **62**, pp. 670–677

37 BONNIN, K. D., and KADAR-KALLEN, M. A.: 'Theory of the light-force technique for measuring polarisabilities', *Physical Review A*, 1993, **47**, pp. 944–960

38 EKSTROM, C. R., SCHMIEDMAYER, J., CHAPMAN, M. S., HAMMOND, T. D., and PRITCHARD, D. E.: 'Measurement of the electric polarisability of sodium with an atom interferometer', *Physical Review A*, 1995, **51**, pp. 3883–3888

39 OSMAN, J., LIM, S.-C., and TILLEY, D. R.: 'Nonlinear optic coefficients in the ferroelectric phase', *Journal of the Korean Physical Society*, **32**, pp. S446–S449

40 ASHKIN, A., BJORKHOLM, J. E., and CHU, S.: 'Caught in a trap', *Nature*, 1986, **323**, p. 585

41 CHU, S., BJORKHOLM, J. E., ASHKIN, A., and CABLE, A.: 'Experimental observations of optically trapped atoms', *Physical Review Letters*, 1986, **57**, pp. 314–317

42 PRODAN, J., MIGDALL, A., PHILLIPS, W. D., SO, I., METCALF, H., and DALIBARD, J.: 'Stopping atoms with laser-light', *Physical Review Letters*, 1985, **54**, pp. 992–995

43 DALIBARD, J., and COHEN-TANNOUDJI, C.: 'Laser cooling below the Doppler limit by polarization gradients, simple theoretical models', *Journal of the Optical Society of America B – Optical Physics*, **6**, pp. 2023–2045

44 RAAB, E. L., PRENTISS, M., CABLE, A., CHU, S., and PRITCHARD, D. E.: 'Trapping of neutral sodium atoms with radiation pressure', *Physical Review Letters*, 1987, **59**, pp. 2631–2634

45 ASHKIN, A., DZIEDZIC, J. M., BJORKHOLM, J. E., and CHU, S.: 'Observations of a single-beam gradient force optical trap for dielectric particles', *Optics Letters*, 1986, **11**, pp. 288–290

46 SMITH, S. B., CUI, Y. J., and BUSTAMANTE, C.: 'Overstretching B-DNA: The elastic response of individual double-stranded and single-stranded DNA molecules', *Science*, 1996, **271**, pp. 795–799

47 AFZAL, R. S., and TREACY, E. B.: 'Optical tweezers using a diode laser', *Review of Scientific Instruments*, 1992, **63**, pp. 2157–2163

48 CELOTTA, R. J., GUPTA, R., SCHOLTEN, R. E., and McCLELLAND, J. J.: 'Nanostructure fabrication via laser-focused atomic deposition', *Journal of Applied Physics*, 1996, **79**, pp. 6079–6083

49 BERGGREN, K. K., BARD, A., WILBUR, J. L., GILLASPY, J. D., HELG, A. G., McCLELLAND, J. J., ROLSTON, S. L., PHILLIPS, W. D., PRENTISS, M., and WHITESIDES, G. M.: 'Microlithography by using neutral metastable atoms and self-assembled monolayers', *Science*, 1995, **269**, pp. 1255–1257

50 BINNIG, G., QUATE, C. F., and GERBER, C.: 'Atomic force microscope', *Physical Review Letters*, 1986, **56**, pp. 930–933

51 WINNEWISSER, C., JEPSEN, P. H., SCHALL, M., SCHYJA, V., and HELM, H.: 'Electro-optic detection in $LiTaO_3$, $LiNbO_3$ and ZnTe', *Applied Physics Letters*, 1997, **70**, pp. 3069–3071

52 JEPSEN, P. H., WINNEWISSER, C., SCHALL, M., SCHYJA, V., KEIDING, S. R. and HELM, H.: 'Detection of THz pulses by phase retardation in lithium tantalate', *Physical Review E*, 1996, **53**, p. R3052

53 PFEIFER, T., HEILIGER, H. M., LOFFLER, T., OHLHOFF, C., MEYER, C., LUPKE, G., ROSKOS, H. G., and KURZ, H.: 'Optoelectronic characterization of ultrafast electric devices: Measurement techniques and applications', *IEEE Journal of Selected Topics in Quantum Electronics*, 1996, **2**, pp. 586–604

54 JIANG, Z., SUN, F. G., and ZHANG, X.-C.: 'Spatio-temporal imaging of THz pulses'. Sixth IEEE International Conference on *Terahertz electronics* 1998, pp. 94–97

55 NOZOKIDO, T., MINAMIDE, H., JONGSUCK, B., FUJII, T., ITO, M. and MIZUNO, K.: 'Visualisation of photo-excited free carriers with a scanning near-field millimeter wave microscope'. 23rd international conference on *Infrared and millimetre waves*, 1998, pp. 382–384

56 BOWEN, J. W., KARATZAS, L. S., TOWLSON, B. M., CRONIN, N. J., BROWN, D. A., WOOTTON, S., AGBOR, N., CHAMBERLAIN, J. M., PARKHURST, G. M., DIGBY, J., HENINI, M., COLLINS, C., POLLARD, R. D., MILES, R. E., STEENSON, D. P., and THOMPSON, D.: 'Micro-machined integrated components for terahertz frequencies'. Proceedings of the 30th ESLAB symposium on *Submillimetre and far-infrared space instrumentation*, *ESTEC, Noordwijk, The Netherlands*, ESA SP-388, 1996, pp. 183–186

57 CRONIN, N.J.: 'Microwave and optical waveguides' (Institute of Physics Publishing, Bristol and Philadelphia, 1995)

58 SNYDER, A. W., and LOVE, J. D.: 'Optical waveguide theory' (Chapman & Hall, London, 1983)

59 ISHIGURE, T., SATOH, M., TAKANASHI, O., NIHEI, E., NYU, T., YAMAZAKI, S., and KOIKE, Y.: 'Formation of the refractive index profile in the graded index polymer optical fiber for gigabit data transmission', *Journal of Lightwave Technology*, 1997, **15**, pp. 2095–2100

60 ISHIGURE, T., NIHEI, E., KOIKE, Y., FORBES, C. E., LANIEVE, E., STRAFF, A., and DECKERS, H. A.: 'Large core high-bandwidth polymer optical fibre for near infrared use', *IEEE Photonics Technology Letters*, 1995, **7**, pp. 403–405

61 IRIE, S., and NISHIGUCHI, M.: 'Development of the heat resistant plastic optical fiber', in Third International Conference on *Plastic optical fibres and applications, POF '94'*. The European Institute of Communications and Networks, (EICN) Yokohama, Japan, 1994, p. 88

62   KOBAYASHI, T., NAKATSUKA, S., IWAFUJI, T., KURIKI, K., IMAI, N., NAKAMOTO, T., CLAUDE, C. D., SASAKI, K., KOIKE, Y., and OKAMOTO, Y.: 'Fabrication and super-fluorescence of rare-earth chelate-doped graded index polymer optical fibers', *Applied Physics Letters*, 1997, **71**, pp. 2421–2423

Chapter 3

Recent developments on amplitude, wavelength, phase and polarisation modulation sensors

S. Hadjiloucas and D. A. Keating

3.1 Overview

In the following sections some recent advances on amplitude, wavelength, phase and polarisation modulating sensors are discussed. Although such division is general, as some recent techniques use many modulation principles simultaneously to separate efficiently the coupled parameters of interest, it still provides a basic framework for the design engineer to choose which technology may be the most appropriate for the solution of a particular problem.

In cases where optical fibres rather than free-space propagating beams are used, optical fibre sensors may be divided into extrinsic or intrinsic ones. This is dependent on whether the light signal is modulated inside or outside the fibre.

Examples of extrinsic fibre optic sensors include encoder plates or disks that measure linear or angular position, reflection and transmission configurations to measure pressure, flow and damage, total internal reflection sensors, for measuring pressure and liquid level, grating based sensors for measuring pressure and vibration, fluorescence sensors for temperature, viscosity and chemical analysis applications, evanescent sensors for monitoring temperature and strain, laser Doppler velocimetry sensors for flow measurement, pyrometers, for sensing temperature, and sensors based on photoelastic effects for monitoring pressure, acceleration, vibration and rotary position.

Examples of intrinsic fibre optic sensors include microbend sensors for measuring strain, pressure and vibration, blackbody sensors for measuring temperature, and distributed sensors. The latter can be further divided into Rayleigh sensors for measuring strain and temperature,

Raman sensors for measuring temperature, mode-coupling sensors for measuring strain, pressure and temperature and quasi-distributed sensors for measuring acceleration, strain, magnetic field and temperature. A further sub-class of intrinsic sensors is that of interferometric sensors, although interferometric sensors can also be implemented in free space as described in Chapter 2.

The most commonly employed interferometric fibre optic sensors are the Michelson system, which is suitable for measuring acoustic waves, electric and magnetic fields, temperature and strain, the Mach–Zehnder configuration for measuring any of the above parameters including current, the Sagnac which also offers rotation and wavelength measurement capability, the ring resonator for rotation and acceleration sensing and the Fabry–Perot configuration. The latter can be implemented in a single-mode fashion for measuring acoustic waves, temperature and pressure or in a multimode fashion for measuring temperature, pressure index of refraction, etc. Polarisation modulation sensors are most commonly employed using a Michelson or Mach–Zehnder configuration and are well suited in measuring acceleration, pressure, temperature and strain.

Although there has been incremental improvement in the performance of most sensors due to recent advances at the component level it is believed that, in the future, many important advances will arise at the design level. Emphasis is therefore given to both devices and systems where the sensor performance can be significantly improved using feedback principles. Furthermore, in order to enhance current possibilities with existing sensing systems, some advances in other parts of the electromagnetic spectrum have been included. As the subject of optical fibre sensors is now more than 20 years old, and most of the sensor schemes have been extensively covered already in many books (Section 3.7), a more selective coverage is provided to inform the reader of some recent developments and to provide a more in-depth description of some selected interdisciplinary topics.

3.2 Amplitude modulating sensors

Intensity modulated fibre optic sensors have been widely adopted by the industry because of their simplicity in operation. Although they are not as accurate as interferometric sensors they offer sufficient accuracy for several process control applications. As displacement sensors, they can make good velocimeters, accelerometers, force and pressure transducers. Slight modification to these sensors allows their application to vibration, axial motion, proximity, rotation, part gauging, film thickness, eccentricity and liquid level sensing. Alternatively, amplitude modulating sensors may be used for chemical, biochemical and environmental sensing.

As mentioned in Chapter 2, the sensors can be extrinsic (whereby amplitude modulation takes place outside the fibre) or intrinsic (whereby amplitude modulation takes place inside the fibre). Typical examples of the extrinsic type of sensors are the reflection and transmission sensors, whereas typical examples of intrinsic sensors are the microbend and macrobend sensors. For extrinsic type sensors, the amplitude is generally modulated by changing the degree of coupling between the emitting and receiving fibres, so that the absorption, emission or the changes in refractive index in the coupling region may be monitored. Using these basic parameters, fibre optics can be used for measuring pressure and strain, vibration, liquid and solid levels, gas presence, temperature, etc., depending upon the transduction method of choice. However, one of the simplest applications of fibre sensors as sensory devices is in position detection.

3.2.1 Reflectance and transmittance based amplitude modulating displacement sensors

In its simplest configuration, a reflectance displacement transducer (Figure 3.1), consists of two fibres (emitter and detector), placed next to each other.

The fraction of light returning to the detector is directly dependent on the sensor gap, that is the fibre-to-reflector separation. Assuming uniform light distribution within the core of the fibre, the fraction of the light detected (i.e. the ratio of the outgoing light intensity to incoming light intensity) for the third case shown in Figure 3.1 is given by [1]

$$P = \frac{r^2}{(r + 2d\tan\theta_a)^2} \tag{1}$$

where r is the fibre radius and d is the target distance. For the first and second cases in Figure 3.1, initially there is a small non-linear region, then a very steep and linear part (the region used for measurements) and finally

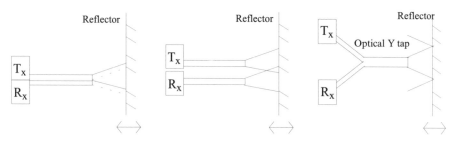

Figure 3.1   Emitter–detector pair amplitude modulating displacement transducers

there is a decreasing region according to 1/(distance)² as shown in Figure 3.2 [2, 3]. The amount of light that is finally coupled into the return fibre is determined by the overlap area of the cone from the image of the input fibre with that of the output fibre.

In many industrial applications, however, fibre bundles are used instead of the already described emitter–detector pair. Krohn [3] showed that it is possible to alter the responsivity and dynamic range of the reflectance sensor by using coaxial, hemispherical or random probe configurations (Figure 3.3). He also showed that the coaxial probe is more sensitive (by a factor of three) than the fibre pair probe and that the fibre pair probe has the largest dynamic range (about 100 mm) over the linear part of the sensor response.

When glass fibres are used instead of plastic ones, many fibres can be packed together in a hexagonal close-packed structure to maximise the diameter of the sensor. In such cases, in order to improve on the coupling efficiency each fibre must be as thin as possible. Toba *et al.* [4] used 3000 fibres, each 14 μm thick, to create a 1 mm bundle. Using a circular bundle of optical fibres, Hu and Cuomo [5] measured diaphragm displacement. Although Fujii *et al.* [6] formulated a model for the power distribution in such a fibre arrangement, the analysis on light intensity functions of the reflectance type sensor provided by He and Cuomo [7, 8] and the analysis by Cook and Hamm [9] are the most comprehensive ones. In some other cases, when a whole reflecting surface must be mapped, it is beneficial to use a surface covered fully with emitting and detecting fibres, and to use a video camera to detect the signal [10].

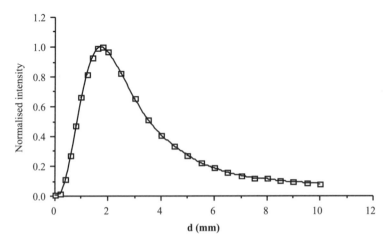

Figure 3.2   Normalised output intensity against distance for an extrinsic fibre pair reflectance sensor

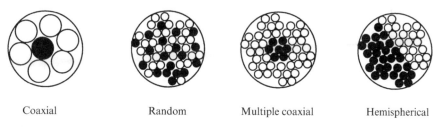

| Coaxial | Random | Multiple coaxial | Hemispherical |

Figure 3.3 *Common fibre bundle emitter–receiver configurations for amplitude modulated extrinsic type sensors*

In order to improve the responsivity of a transducer based on PMMA fibres, a simple way is to cut the emitter (T_x) and receiver (R_x) fibres at an angle a so as to make full use of the critical angle. This ensures improved coupling, and therefore a better signal-to-noise ratio. The condition for a is the one for which the ray emerging from the fibre is parallel to its sloping face, and using Snell's law to cut the fibres gives $n_o\sin90° = n_l\sin(17.63° + a)$, so $a = 24.52°$. This value is used for the reflectance displacement transducer shown in Figure 3.4b, where the fibres are placed next to each other. A further improvement in performance can be obtained by cutting the fibres twice, at angles of 24° and 66° as shown in Figure 3.4c. The only dual fibre optic sensor which investigates the effect of varying the angle between the two fibres has been described by Powell [11], where an increase in sensitivity of only one order of magnitude is reported. The advantage of the double-cut configuration is that it maximises the amount of the reflected light that enters the receiving fibre, and minimises the reflected light that is lost in all other directions. This results in an increase in responsivity at the expense of range (Figure 3.5). The collective results for responsivity, SNR, least detectable signal (LDS) and linear range are shown in Table 3.1.

The response function of the double cut transducer is proportional to the flux entering the receiving fibre. This can be analysed as a flux transfer problem. Assuming a perfect reflector the receiving fibre can be thought to be facing the emitting fibre at a distance twice that between the emitter fibre tip and the reflector (Figure 3.6). The fibre end face is cut at an angle θ_{cut} with respect to its axis and the emitting light is re-directed. The face of the cut fibre has an elliptical shape with minor axis r_f (the fibre radius) and major axis $r_f' = r_f/\cos(\theta_{cut})$. The ray inside the emitting fibre incident on the cut surface at an angle θ will emerge at an angle given by the bend function $B(\theta) = \sin^{-1}(n\sin\theta)$ according to Snell's law, where n is the refractive index of the fibre, with n_{air} taken as 1.

The radiance at a point (x, y) in the fibre and at direction $(\theta\ \phi)$, is given by $L(x,y,\theta,\phi)$. In order to solve the flux transfer problem the

Table 3.1   Measured values of responsivity, signal-to-noise ratio, least-detectable signal and linear range, for the different sensors at unit bandwidth

Sensor type	Responsivity (V/m)	SNR (dB)	LDS (m)	Linear range (m)
Uncut	183	62.5	4.3×10^{-10}	1200×10^{-6}
Single cut	770	64.6	1.4×10^{-10}	650×10^{-6}
Double cut	11 880	67.8	1.7×10^{-11}	80×10^{-6}

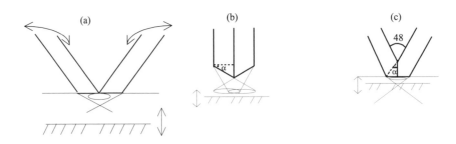

Figure 3.4   Different configurations of fibre optic pairs showing (a) the original configuration suggested by Powell for glass fibres, (b) optimal cutting suitable for PMMA fibres at 24 degrees and (c) double cut configuration with cuts at 24 and 66 degrees

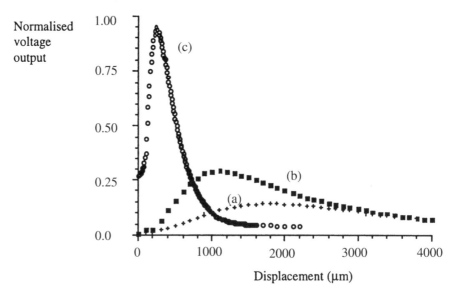

Figure 3.5   Response curves for (a) uncut, (b) single cut and (c) double cut fibres, using a mirror reflector

radiance is integrated over the transmitting area S and the receiving area A:

$$\Phi = \int_S \int_A \frac{L' \cos\theta' \cos\psi}{R^2} \, ds \, da \qquad (2)$$

where L' is the radiance function at the surface of the transmitting fibre, θ' is the exit angle of the ray from the transmitting surface, R is the range of the receiving point in relation to the transmitting point, and ψ is the incidence angle of the ray on the receiving surface which is, in fact, equal to the exit angle θ' from the transmitting surface as the two surfaces are parallel. A similar observation can be made for the azimuth of the ray from the transmitter surface so that $\phi = \omega$.

The radiance L' is derived from the function L evaluated at the

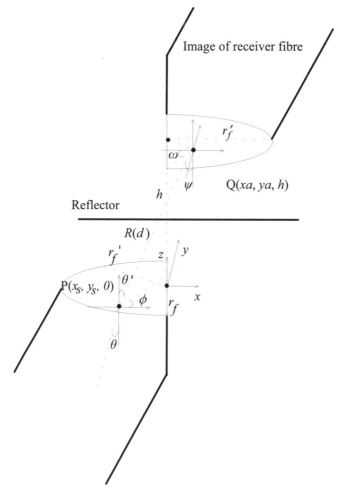

Figure 3.6   Diagram of light coupling between two double cut fibres

position of the source area element ds and at a direction (θ, ϕ) such that $B(\theta) = \theta'$ and θ' and ϕ are calculated from the relative positions of the transmitting and receiving surface elements ds and da.

In Figure 3.6 we first assume a point source P on the emitting cut fibre with co-ordinates $P \equiv (x_s, y_a, 0)$. A point Q at position (x_a, y_a) relative to the position of the centre of the receiving surface will have co-ordinates $q \equiv (x_a, y_a, h(d))$ and $h(d) = 2d$. The distance R can be calculated from:

$$R(x_s, y_s, x_a, y_a, d) = |\mathbf{PQ}| = \sqrt{(x_a - x_s)^2 + (y_a - y_s)^2 + h^2} \tag{3}$$

The flux incident on the reflector surface results in a second source that is at a distance d from the emitting and receiving fibres. For measurements when the modulation observed is not related to displacement this distance may be optimised in each particular application for minimum standing noise at the photodetector. Integration is then performed at a fixed distance for all the points on the reflector and all the points on the receiving fibre after taking into account an attenuation function $\mathfrak{I}$ that is related to the physical process observed. The angles θ' and ϕ are calculated from:

$$\phi(x_s, y_s, x_a, y_a) = \tan^{-1}\left[\frac{y_a - y_s}{x_a - x_s}\right] \tag{4}$$

and

$$\theta'(x_s, y_s, x_a, y_a, d_{opt}) = \cos^{-1}\left[\frac{2d}{R(x_s, y_s, x_a, y_a, d)}\right] \tag{5}$$

In order to calculate θ we can use $B^{-1}(\theta')$ so that $\theta = \sin^{-1}(\frac{1}{n}\sin\theta')$. The flux on the receiving fibre is therefore:

$$\Phi(d_{opt}) = \int_{-r_f}^{r_f}\int_0^{r_f'}\int_{-r_f}^{r_f}\int_{-r_f'}^0 \mathfrak{I}\frac{L(x_s, y_s, \theta, \phi)\cos^2\theta'}{R^2(x_s, y_s, x_a, y_a, d)}\,dx_s dy_s dx_a dy_a \tag{6}$$

In many applications, when an expanded dynamic range is required, a lens system in conjunction with a fibre optic probe may also be used. For collimation purposes a special kind of lens called the graded index (GRIN) rod lens may be used. Using this lens at the fibre end face is generally advantageous since the focal point of the sensor, when coincident with the fibre end face, may provide easier and superior coupling (Figure 3.7).

Tests on misalignment losses between GRIN rod lens couplers indicate that such losses are largely independent of the linking fibre lengths. This contrasts with direct fibre–fibre coupling losses which are dependent on the modal power distribution in both the feed and return fibres. GRIN lenses have therefore been used in many sensor designs [12, 13]. The advantages of a GRIN rod lens in a measuring system in terms of linearity and responsivity have been highlighted by Cusworth and Senior

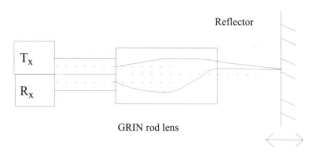

Figure 3.7   Diagram showing the operation of a GRIN rod lens at the end of a fibre pair

[14]. An alternative to the GRIN lens is the Selfoc lens, and this may also be used to improve the coupling in a fibre emitter–detector pair [15].

When the reflectivity of a material is low and the surface irregularities are few, so that the light beam may travel through a fixed path length, transmittance measurements are possible. A commonly employed technique for position sensing is the use of a filter that constantly changes its transmittance as a function of position with respect to the emitting and receiving fibres. In its simplest form the technique may employ samples of computer generated positive film prints that have a linear or exponential dot pattern along their length [16].

For better resolution and dynamic range, however, masks are employed. In a mask motion sensor, the input fibre is imaged by a two lens system on the output fibre and the mask travels perpendicularly to the direction of light propagation between the two lenses. If higher sensitivity is desired, two periodic masks consisting of alternate transparent and opaque regions of equal width may be used. This forces the transmission through the pair of masks to vary from 50%, when the masks completely overlap, to zero when the opaque bars of one are completely over the transparent regions of the other (Figure 3.8).

This technique is most commonly known as Schlieren intensity modulation and has been exploited by Spillman and Mahon [17], to measure successfully pressure fluctuations on a diaphragm.

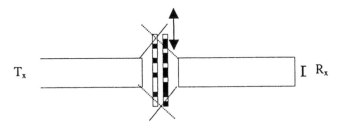

Figure 3.8   Intensity modulation using two periodic masks

In some other applications, as an alternative to a mask, a moving refractive microsphere, a lens or a ferroelectric liquid crystal spatial light modulator that changes between transparent and opaque in the presence of an electrical signal can also be used to modulate the intensity of light coupled. Alternatively, using a shaft encoder between an emitting and a receiving fibre, rotation can be sensed; such measurements can be further extended to flow measurements by simply attaching fins to the encoder. Using an array of sources and detectors and with suitable patterns (binary or grey code) on the encoder, a digital signal which is proportional to rotation can be obtained directly.

Finally, there are many other sources of splicing loss in fibres that can be exploited to make an amplitude modulated displacement transducer. These are reflection loss at the end surface of the fibre, lateral or longitudinal offset of fibres, tilt, deformation of end surfaces, and differences in fibre parameters such as core-cladding diameter, numerical aperture, or fibre refractive index profile (Figure 3.9).

Perhaps the most striking example of the above mentioned sensor configurations has been the use of a lateral displacement sensor [18] where two silica glass fibres are separated by about 2–3 μm and the degree of coupling is measured in an enclosed cell (hydrophone). The technique claims a minimum resolution of 8.3×10^{-3} Å (Figure 3.10).

Although many techniques in the literature quote very small minimum resolutions, in practice these are not achieved due to electronic noise and drift. The latter, however, may be minimised using appropriate intensity referencing schemes (described extensively in the article by Murzata and Senior [19]).

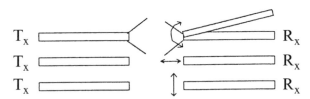

Figure 3.9   Angular, longitudinal and transverse configurations for displacement sensing

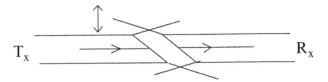

Figure 3.10   Lateral displacement sensor of improved resolution as suggested in [18]

3.2.2 Intensity referencing schemes

As with all other sensors and transducers, short and long term drift of the amplitude modulating sensors are of paramount importance. Two general methods can be applied to maintain calibration of the instrument, either by sweeping through its full dynamic range, or by providing a reference signal with which to compare the sensing signal.

The compensation of these changes, usually seen as long term drift, is therefore a major concern when designing an intensity-based sensor system, and a considerable amount of research has been conducted in this area in the past few years. In 1983, Culshaw [20] demonstrated a compensation optical bridge concept using two light sources operating at the same wavelengths. A more elaborate loss-compensation technique (Figure 3.11) has been described by Wang *et al.* [21].

There are many reference schemes that one may choose from: self-referencing schemes, dual receiver single emitter schemes, or dual wavelength intensity referencing schemes.

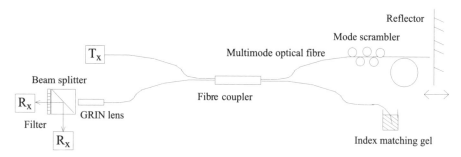

Figure 3.11   Compensation scheme proposed by Wang et al. [21]

3.2.2.1 Self-referencing schemes

In extrinsic type sensors, it is possible to use two partially reflecting mirrors at the ends of a fibre microbend sensor, thus forcing the light pulses to travel many times backwards and forwards through the sensing element so that the cumulative signature has an increased signal-to-noise ratio. This has the further advantage that taking the ratio of two pulses results in a strong rejection of fluctuations in coupling [22].

In a similar manner, in extrinsic type sensors, when multiplexing schemes are used, self-referencing of a sensor can be achieved using a sensor loop [23]. When monitoring a short pulse recirculating within the loop, the first pulse received by the detector will be proportional to the intensity of the light coupled into the loop, the second pulse will be proportional to the intensity modulation, the third pulse will be proportional to the intensity modulation squared and so on (Figure 3.12). The high attenuation coefficient of plastic optical fibres precludes their use in systems where such referencing schemes must be used.

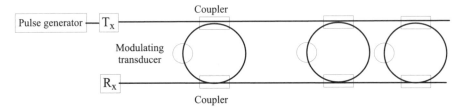

Figure 3.12   Self-referencing scheme for multiplexed amplitude modulating sensors

3.2.2.2 Dual receiver single emitter schemes

An interesting variation of the simple reflective probe is the use of a dual probe with a common light source to provide some source of referencing. The dual output can normalise light-source intensity variations as well as increase the output for a given movement, thereby increasing the accuracy. In addition to magnitude, such configurations can provide directional information if one probe is working on the front slope, whereas the other is working on the back slope. The advantage is that the sensing probes can be placed on one side of the object and no longer have to oppose one another (Figure 3.13).

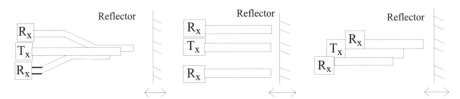

Figure 3.13   Dual probe fibre displacement sensors

In order to provide compensation due to changes in the refractive index of the environment (chemical applications) or target reflectivity, a second sensing fibre can be introduced [24]. A similar configuration has also been suggested by Libo and Anping [25] with a small difference in the spatial configuration of the probe. A very interesting technique for linear scaleable optical sensing of displacement has been suggested by Kalymnios [26]. The technique utilises one emitting fibre and two receiving fibres which are positioned at different distances with respect to the emitter so that the ratio of the sum and difference between the two signals is used.

Alternatively, when directional information must be provided but there is only one photodetector, a system of two LEDs modulated with a square wave in antiphase, together with a movable aperture and a fixed aperture, can be used instead [27]. When centred, the movable aperture receives an equal amount of light from the two fixed apertures and thus from the two LEDs (Figure 3.14). As it moves from one side to the other, the light from either the first or the second source will dominate. Since the LEDs are modulated with opposite phases, the phase of the ac component of the detected signal will change by 180 degrees on either side of the centre. The ac signal which is proportional to the displacement is extracted by a high pass filter and then converted to dc (demodulated) using a phase sensitive detector.

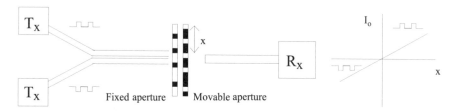

Figure 3.14   Set-up and signal outputs for sensing the movement of an aperture with respect to a fixed aperture

In another application, using a three probe arrangement in which two probes illuminate an opaque material at an angle and a third vertical sensing fibre is positioned between the other two, Borsboom and TenBosch [28] have monitored the reflected and scattered signal from dental enamels. However, other applications such as quality control of translucent materials, and scattering from plant leaf tissues (detecting the opening and closing of stomata during transpiration), is also possible.

3.2.2.3 Dual wavelength intensity referencing schemes

Using flattened end fibres, a two colour optical displacement sensor incorporating an intensity referencing scheme can be realised using transmission or reflection type colour filters in many configurations, as in Figure 3.15.

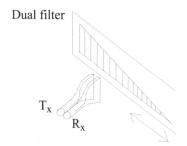

Figure 3.15   Flattened end fibres and two colour optical displacement sensor using reflection type colour filters

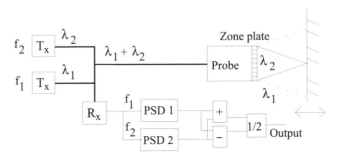

Figure 3.16   Two wavelength sensor using a zone plate

Alternative designs based on a focused beam reflective principle have also been suggested by Jones and Spooncer [29]. The displacement sensors described therein utilise a two-wavelength operation using a zone plate. The output characteristics are resolved by difference/sum referencing (Figure 3.16). Dual colour pulsed LEDs that could be used for this purpose are commercially available nowadays. The two wavelength signals can also be obtained from a single broad-band LED source by dividing its spectral emission into two semi-broad-band signals so that the wavelength bands associated with each signal lie on opposite sides of the LED peak wavelength.

Such a scheme, however, can lead to serious measurement errors with variations of the LED operating temperature. LED instability may originate from changing ambient conditions or from changes in drive current. For an LED, an increase in temperature can lead to a diminishing of the optical power output, a spectral shift of the peak wavelength which is generally towards the longer wavelengths, and an increase in spectral bandwidth.

An advantage of the dual wavelength referencing technique is that both wavelength signals can be transmitted in parallel through the system by employing wavelength division multiplexing components. However, this provision alone cannot guarantee that both wavelengths will behave identically to all perturbations as is sometimes assumed. The wavelength bands must still be selected to minimise the differential intensity variations between the two signals arising from unequal spectral transmission effects. However, these differential variations are difficult to predict, particularly when the sensor system incorporates multiple passive optical components. In this case, the measurand and reference signal wavelength bands should be closely spaced to ensure similar behaviour with all spectral transmission effects. Unfortunately, the parallel transmission of signals with closely spaced centre wavelengths through an optical fibre gives rise to optical crosstalk which then dictates the overall measurement accuracy of the sensor.

3.2.3 *Examples of amplitude modulating sensor applications*

3.2.3.1 *Chemical and biochemical sensors*

Apart from displacement type sensors, and based on the principles demonstrated earlier, a range of other sensors may be realised using optical fibres (glass or plastic type). Some examples of chemical, biochemical and environmental sensors follow.

Provided that part of the optical fibre comprising the sensing element can be made porous, some chemicals of interest can easily diffuse to its core. An indicator that reacts to the chemicals may induce the attenuation in the received signal. Mixing chemically selective indicators in the monomer solution prior to the formation of a plastic fibre usually results in a uniform distribution of the indicator into the polymer matrix. Such sensor characteristics as dynamic range, linearity and response time can be tailored to meet specific applications by altering the polymer composition and polymerisation procedure. The amount of the indicator immobilised in the sensor can be precisely controlled, resulting in excellent sensor-to-sensor repeatability.

Another important parameter that can be exploited using amplitude modulating sensors is refractive index. A technique that exploits the change in the critical angle and thus a change in internal reflectivity of the fibre has been proposed by Spillman [30] to measure refractive index (and hence concentration) changes in a liquid (Figure 3.17). Such a scheme eliminates the problems of compensation required by extrinsic type sensors where the received amplitude at the detector is a function of both refractive index and concentration.

Alternatively, an intrinsic optical fibre refractometer that uses attenuation of the cladding modes can be constructed. Several liquid level sensors based on such principles have been described. Apart from refractive index measurements, it is also possible to perform measurements of aqueous kinetics [31] at micromolar concentrations in a range of solutions. Alternatively, molecular weight sensing using flow injection analysis and refractive index gradient detection is possible [32].

By introducing new additives (dyes) in the fabrication processes, highly efficient fluorescent plastic optical fibres can be constructed at a range of diameters from 0.2 mm to 5 mm. These are able to absorb ambient light incident along their entire length and emit it at longer wavelengths,

Figure 3.17   *Amplitude modulating sensor of improved responsivity for the determination of refractive index of solutions*

therefore making a very sensitive ambient light detector [33]. The same fibres can also form the basis for a scintillating sensor that converts nuclear radiation (X-rays or gamma rays) into blue light [34]. However, this requires the conversion of blue light into green light before finally being detected by a photomultiplier.

With new fabrication processes, other shapes such as square, rectangular and even hexagonal sections can be realised in the transverse or longitudinal direction of the fibres [35]. As such, these fibres can be used in large video screens for continuous monitoring of processes. Perhaps the greatest advantage of such fibres, however, is their ability to focus the collected light signal from their surfaces on a smaller area photodetector, increasing the signal-to-noise ratio.

Indicator dyes have also been used to form an ammonia sensor just by modifying the porous fibre itself, making it highly gas permeable but water and ion impermeable. Alternatively, using palladium chloride as an indicator, a carbon monoxide sensor can be produced, as the latter reacts with the former forming metallic palladium (a black film that attenuates the light); however, the reaction is irreversible. Generally, a dual wavelength intensity referencing technique is usually preferred in this sort of sensing to minimise variations in the connector attenuation, fibre bending, the detectors' response, biofouling and indicator loss. Recent advances in chemical sensors can be found in the article by Wolfbeis [36].

An area of considerable interest for chemical sensing is that of evanescent field sensors [37–39]. For light incident at the interface between two optically transparent materials of high and low refractive index, an interaction between incident and reflected light occurs, resulting in a standing wave close to the interface. This standing wave decays exponentially away from the interface into the low refractive index material (Figure 3.18).

The depth of penetration of evanescent fields is approximately around $\lambda/5$ when normal incidence light is used, and takes its maximum value of λ at about one degree greater than the critical angle. The simplest form of interaction with the evanescent field is absorption, although the phenomenon has also been used in commercially available internal reflection

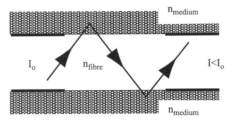

Figure 3.18   Basic principle of operation of amplitude modulation evanescent field sensor

spectrometers operating at near-infra-red wavelengths. More recently, evanescent fields have been used for the excitation of fluorescence, and for the monitoring of ligands to layers of antibody material immobilised on the surfaces of waveguides. Molecular detection sensitivities of around 1 nM have been demonstrated using this technique and the technique holds considerable promise for antibody–antigen studies. Furthermore, by covering the interface between the two transparent media with a thin metal layer, surface plasmon resonance is obtained, and a variety of sensors exist based on this principle. A recent comprehensive review on evanescent fibre sensors may be found in the article by Stewart and Johnstone [40].

The use of fine optical fibres in conjunction with microscopes and high resolution translational stages offers the possibility to monitor bio-chemical changes at the cellular level. A fine example of optical fibre sensors illustrating this is the pH fibre optic micro-optrode for intracel-lular measurements based on fluorescence described in [41]. An overview of biosensors can be found in [42].

3.2.3.2 Environmental sensors

In the field of environmental sensing, in the far-infra-red, measurements in the 1 μm to 1 mm region provide meteorological information as well as insight on the origin and evolution of our solar system and the formation of galaxies and stars in the early universe. A variety of sensing schemes exist in the infra-red and visible parts of the spectrum, where indicator dyes, polymers and additives deposited on various support materials such as Teflon [43], sol-gels [44], plastic tips, fibres, integrated waveguides or walls of disposable cuvettes and inside capillaries, may be used for sensing pH [45], oxygen, carbon dioxide and certain ions. The evaluation of appropriate porous glass materials as a support for sensitive reagents for chemical sensing is currently a topic of active research. Simple opto-electronic schemes may often be sufficient for these sensors. A recent example includes the monitoring of ozone using the strong absorption band of ozone (600 nm), using a modulated LED source [46]. Generally, for monitoring the small changes in optical absorption (commonly encountered in environmental sensing), an absorption cell must be used. A technique for monitoring very small changes in absorption using a double beam photometer and two photomultiplier tubes has been described by Mitchell and Wayne [47]. In the UV region, examination of nitrate pollutants using a water-core waveguide is also possible [48] and the total organic carbon (TOC) in water sources and supplies may be monitored using a variety of techniques [49] although, occasionally, complicated processing schemes using neural networks might have to be implemented [50]. A detailed overview of environmental sensors can be found in [51].

The double cut configuration described previously when aligned at its optimal distance (point of inflexion of the response curve) may further be implemented to measure relative humidity. This is achieved by optically monitoring dew formation on a Peltier cooled mirror surface. Once formation of dew is detected by the optical sensor, the current through the Peltier element is reversed until the mirror reflectance increases above a pre-set reference value (Figure 3.19).

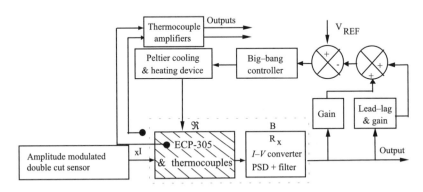

Figure 3.19   Block diagram of a feedback dew-point sensor [52]. The system uses a fibre-optic coupler to re-direct some of the emitted light (1-x) for intensity stabilisation of the LED.

Operating the sensor about the point of dew formation ensures that the mirror temperature is equal to the dew temperature (the difference between the two being the error signal in the feedback loop). Rapid response thermocouples connected to cold junction compensated thermocouple amplifiers may be used for measuring mirror and ambient temperatures so that the relative humidity may be calculated.

3.2.3.3 Temperature and pressure sensors based on amplitude modulation

Amplitude modulation techniques utilising optical fibres can also be used for sensing temperature [53]. Augousti *et al.* [54] observed that the degree of reflection of the red to infra-red ratio on a liquid crystal surface is highly dependent on temperature. Similarly, Kajanto and Fridberg [55] showed that the temperature dependent light absorption properties of semiconductors (which depend on the band-gap energy between valence and conduction band) could also be used for a broad range temperature sensor. Using the same principle, an LED can be used as a dynamic pressure transducer [56], by monitoring the change of forward bias voltage with applied pressure.

Fluorescence based optical thermometers have been reported to be capable of measuring temperatures up to 1250°C [57]. A comparison of the performance of fluorescence intensity ratio and fluorescence lifetime schemes can be found in [58]. If the response time of a fluorescence sensor is not an issue, it is possible to use porous glass as a support to the sensitive reagent so as to further amplify the fluorescence signal [43].

3.2.3.4 Photo-acoustic resonators

A variety of sensors may also be implemented to measure liquid level, mass flow, wind speed, density, pressure and force, using photo-acoustic resonators. Resonating elements are driven optically using radiation pressure, optothermal phenomena or electronically induced mechanical strain, in which light absorbed by an absorber, deposited on a vibrating element, causes a localised heating effect which in turn induces a thermal stress and therefore a bending force in the element. The efficiency of conversion from optical energy to mechanical energy is linearly dependent on the temperature differences which can be achieved as a consequence of optical absorption and, therefore, the displacement of an optically energised vibrating structure is directly proportional to the optical power used to drive it. By removing the light the element is allowed to relax, and it is possible to modulate the light so as to excite the element at its resonance frequency using feedback. The energy for maintaining the element in the resonant mode is so small that it can be provided by an LED, and the resonance amplitude can be detected with an optical fibre displacement transducer (using only amplitude information or interferometrically). This information can then be used to modulate the LED, thus implementing an amplitude controlled feedback loop.

Other resonant elements, such as piezo-electric quartz, are possible alternatives to the wire element, particularly as their Q value is very high and the frequency maintaining energy is low. By coating the quartz crystal with an absorption layer which can selectively and reversibly absorb or liberate a gas or vapour, the resonant frequency can be altered. Typical quartz vibrating elements configured into a weight sensor [59] are able to measure up to 500 g with a frequency change of 2.0 Hz g^{-1}. Culshaw [60] showed that miniature microresonators (where their size is similar to the size of the optical fibres optically activating them) have response times of a few milliseconds and are capable of measuring temperatures at a resolution of 0.01°C, and pressures to better than 0.01% over a span of 1 bar. Currently, silica double resonant bridge structures are a very active field in sensor technology research, as they can combine measurements of both temperature and pressure on a single sensor chip. An important advantage of microresonators over other sensors is that many of them can be multiplexed using multiple resonant frequencies and a single fibre link.

3.3 Wavelength modulating sensors

The major advantage in using a wavelength modulating technique over amplitude modulation is the immunity of the system components to attenuation in the fibres or connectors. One of the most widely adopted measurement techniques involving wavelength or colour remote sensing is the use of an optical radiation pyrometer. This is simply a sapphire fibre enclosed in a black body cavity whose function is to transmit the reflected broad-band radiation from the cavity to a spectrometer that computes the temperature of the cavity by examining two narrow bands in the received spectrum.

Using a neodymium doped fibre, a differential absorption character-istic with temperature can be observed, and the ratio of the responses at two wavelengths is a function of temperature. Optical components are not always necessary to produce the spectral pattern, and thermo-chromic substances such as cobalt chloride in an isopropyl-alcohol/water solution that changes colour between 25°C and 75°C can be spectro-graphically measured using a two wavelength detector system. However, these systems suffer from long equilibration time constants.

In another technique, light from an LED transmitted by a fibre is absorbed by a GaAlAs crystal, and is then re-emitted at another wavelength (photoluminescence) which is temperature dependent, and analysed by a two detector wavelength system.

Wavelength modulation can also be achieved using white light and a displacement technique, such as by rotating a prism, using a diffraction grating or by linear displacement of the receiving fibre from an emitting fibre followed by an achromatic lens. Alternatively, a zone plate can undergo linear displacement altering the wavelength of the light reflected into a receiving fibre.

Some other wavelength modulating techniques utilise the parameter sensitivity of optical filters, as birefringence, interference and semi-conductor properties are all temperature dependent. Such filters consist of a finely dispersed transparent material in a liquid having a refractive index that varies with temperature. When the indices of the two com-ponents equalise at a unique temperature and wavelength, the filter becomes highly transmissive. Using a broad-band source, the different temperature wavelength pairs at maximum transmission passing through the filter can be measured spectroscopically.

Wavelength modulation can also be achieved through temperature fluctuations due to the wavelength dependence of birefringence in birefringent crystals such as lithium niobate. In this case, a lithium nio-bate crystal sandwiched between crossed polarisers is illuminated with white light and the chromatic transmission function of the system is determined using a monochromator. The modulated intensity signal is related through a sinusoidal function to temperature and is proportional to the difference in refractive index between an ordinary ray and an extraordinary ray, the crystal thickness and the wavelength increment.

Another important class of components in wavelength modulation is Bragg cells (Figure 3.20a). Commonly, Bragg grating structures are photo-inscribed into doped fibres (Ge) using UV light from interference patterns produced either interferometrically or using phase masks. During measurements, the grating may be illuminated using broad-band light from super-luminescent diodes, edge-emitting LEDs or fibre super-fluorescent sources, and the reflected light may be monitored using a spectrometer, a wavelength to amplitude conversion technique using passive optical filters, a tracking technique using a tuneable filter or an interferometric Fourier transform scheme. Typically, Bragg grating sensors operating at 1.3 μm show a change in wavelength of 1 nm per 1000 με. Because of a similar shift of approximately 0.5 nm for a change in temperature of approximately 50°C considerable efforts have been made to discriminate between temperature and strain effects. One recent technique [61] uses wavelength information from the first and second diffraction orders of the grating element. Although such an approach ideally requires the gratings to be deliberately saturated to give stronger harmonics (square wave refractive index profile) separation of the two components is still possible using commercially available gratings. Because the analytical theories of Bragg reflectors and resonant transmission filters are currently well understood [62], it is expected that a range of interesting structures will emerge in the near future.

Similarly to the photo-inscribed optical fibre Bragg grating structures, at the mm-wave part of the spectrum, it is possible to realise two dimensional photonic band-gap Bragg grating structures in microstrip (Figure 3.20b). These are periodic patterns formed by circles etched in the ground plane of the microstrip line [63]. Increasing the width and depth

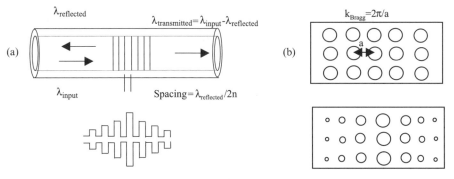

Figure 3.20   *(a) Common Bragg grating sensor inscribed in optical fibres and (b) the equivalent two dimensional photonic band-gap microstrip structure used at mm-wavelengths. Tapering for the realisation of a Hamming window is also shown*

of the stop-band increases the ratio between the circle radius and the period of the structure. This results in an increase in the pass-band ripple which can, however, be suppressed using a tapering technique to the periodic pattern. Recent efforts concentrate on the production of grating patterns produced by optically projecting a grating pattern on to a semiconductor wafer. In regions where light falls the conductivity is high because of the formation of a photo-induced electron–hole plasma, whereas in other areas the conductivity remains low [64]. The aim is to use optically controlled diffraction gratings to control millimetre wave beams propagating in free space.

Another area recently attracting considerable interest is the interaction of Bragg grating structures inscribed in fibres with acoustic waves. Using a standard Bragg cell, a travelling acoustic wave forms a refractive index grating due to the elasto-optic effect. Light of optical frequency ω_0 passing through this region at a specific angle with respect to the acoustic wave propagation direction is diffracted, and its frequency is shifted by the acoustic frequency ω_a. The angle θ of diffraction from the incident angle (where $\sin\theta = \lambda_0/2\lambda_a$, with λ_0 and λ_a being the optical and acoustical wavelengths, respectively) is then isolated by the original carrier using an angle-dependent spatial filter and the final frequency emerging from the filter becomes $\omega_0 + \omega_a$ (with typical operational frequencies ranging from 40 to 1000 MHz). In optical fibres, the acousto-optic interaction can be achieved as long as the carrier at frequency ω_0 and the shifted wave at frequency $\omega_0 + \omega_a$ are guided modes of the fibre. The difference in propagation constant between the carrier and the frequency shifted modes is $\Delta\beta = 2\pi/\lambda_a$, where $\omega_a = 2\pi v_a/\lambda_a$, v_a being the acoustic wave velocity. Some of the limitations of acousto-optic demodulators may be found in the article by Brooks and Reeve [65].

In interactions between optical fibres and acoustic waves, generally, two types of mode are used. These are either the two orthogonal polarisation modes of high birefringence (HB) fibres or the two transverse spatial modes (LP_{01} and LP_{11} modes) of two-mode fibres. Modal filters prior to and after the region where periodic coupling takes place are used to ensure that only the shifted frequency is propagated. When the acoustic wave is travelling in the same direction as that of the optical wave, coupling from the fast mode (smaller propagation constant) to the slow mode takes place, resulting in a frequency up-shift. Alternatively, birefringent fibre frequency shifters can be produced using lateral stress in a direction other than that of the birefringent axis so that power coupling may occur between two originally normal polarisation modes due to the additional stress birefringence. The lateral stress is induced from the interaction of acoustic waves generated on a fused quartz substrate and the fibre, and polarisers are used to ensure polarisation coupling only within the acoustically modulated region.

Two-mode fibre frequency shifters can be produced using the fibre modes LP_{01} and LP_{11} [66], when mode-coupling is achieved by the periodic bending of the fibre. This can simply be induced by a silica acoustic horn excited by a PZT transducer. In this technique, light from a single-mode fibre is launched in a two-mode fibre which, when coiled with a small bending radius, acts as an LP_{01} filter. An LP_{11} filter can be realised using a directional coupler involving a two-mode optical fibre and a single-mode fibre of unity coupling to the LP_{11} mode, and negligible coupling to the LP_{01} mode.

Finally, another important sensing technique that is based on wavelength modulation is laser Doppler velocimetry. The technique uses the Doppler frequency shift which occurs when light is scattered from moving particles at a point location inside a flow stream. The wavelength of the reflected light is decreased or increased according to whether the direction of incident light has a vector component which is in the direction of flow or in the reverse (Figure 3.21). The reflected wavelength λ' is then equal to $\lambda_0(1 - u/c)$ where λ_0 is the wavelength of incident light, and u is the particle velocity relative to the observed direction. Some recent advances on the subject can be found in the articles by Jones and Barton [67] and Lockey and Tatam [68].

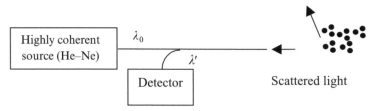

Figure 3.21   General principle of laser Doppler velocimetry

3.4 Phase modulating sensors and interferometric techniques

Optical interferometry is a well established technique which is associated with high resolution optical spectrometry and the precise measurement of length. Apart from length, other important parameters can be derived such as velocity, acceleration, pressure and force. The requirement of good alignment of the optical components constituting the interferometer has prohibited the use of this high resolution technique outside the laboratory in the past. This major restriction of the general application of interferometry has recently been relaxed with the development of all monomode fibre optic equivalents of the classical interferometers. Fibre-optic interferometric measurement systems can be classified as either intrinsic (where the measurement beam always remains within the fibre and the optical properties of the waveguide are changing) or extrinsic (where part of the measurement beam travels outside the wave-guide). Of the intrinsic type, the differential (polarimetric) interfer-ometers can be used for measurements of displacement but suffer from low resolution, unless special fibres are used. There are many extrinsic forms of interferometers but the main ones are the Fabry–Perot, the Mach–Zehnder, the Michelson and the Sagnac. External applications of interferometry include holography, laser velocimetry using double beam interferometry and measurements of displacement and vibration.

Fibre optic modulators are generally used for coherent applications such as interferometry. They provide a means of either stabilising an interferometer against environmental disturbances or of heterodyning to detect phase shift. Phase modulation is achieved by externally modu-lating the length of a fibre or the effective index of the guided mode. Axial elongation, radial stress and temperature are the most common methods to achieve phase modulation, although temperature is not commonly used because of its large time response. Acoustically activated modulators using axial elongation or radial stress such as lead zirconate titanate (PZT) ring devices are one of the commonest devices currently used.

In a PZT modulator, the fibre is wrapped around the piezoelectric tube and a voltage is applied across the cylinder wall, changing its circum-ference and height thereby changing the fibre length. The induced phase shift is given by $\Delta\phi = 2\pi/\lambda(n\Delta l + l\Delta n)$, and for visible light, using modulating frequencies from DC to few kilohertz, a 2π rad phase shift can be realised for an applied voltage turn product of 70 to 100 V turns [69]. The modulation amplitude can be further increased by operating the modulator at one of the acoustic resonance frequencies of the cylin-der. In hollow cylinders with thin walls, the resonance frequency is inversely proportional to the diameter of the cylinder, with the higher frequencies being observed using solid PZT disks. Phase modulation with sawtooth and rectangular waveforms is generally possible only at low frequencies (less than 1 kHz) because of the higher harmonic con-tent of these waveforms. For higher frequencies, fibres with coaxial piezoelectric transducers that upon acoustic excitation laterally squeeze the fibre, modulating its refractive index through the elasto-optic effect, may be used. Well constructed modulators based on the above principle can reach MHz modulation frequencies and are generally free from undesirable polarisation modulation.

Fibres with piezoelectric jackets that respond to an applied electric field modulating the refractive index of the fibre as well as the fibre length are used when broad-band modulation is desired. At such fre-quencies where the extensional acoustic wavelength is longer than the interaction region, both effects contribute to phase modulation, whereas, at higher frequencies, the change in refractive index dominates. These devices have long interaction lengths (longer than the coaxial piezoelectric transducers) and can have a uniform frequency response with a phase modulation of 0.01 rad V^{-1} m^{-1} over a frequency range of 30 kHz to 2.5 MHz [70]. Even higher frequency modulators are possible using a piezoelectric thin film of oriented ZnO along a short section of the fibre. The maximum available phase modulation in these devices is limited by the thermal dissipation and the short interaction length, whereas the fundamental upper frequency limit is determined by the condition that half the acoustic wavelength in the fibre should be longer than the diameter of the optical mode. Although the fibre modulators described offer negligible optical losses and high optical power-handling capabilities, waveguide modulators based on the electro-optic effect in $LiNbO_3$ exhibit a uniform response over an extremely wide range of modulation frequencies and are a possible alternative.

Phase modulation has both advantages and disadvantages. The main advantage is the extremely high degree of sensitivity. Culshaw [71] showed that modulation sensitivities of the order of 100 rad m^{-1} C^{-1} or 10 rad m^{-1} bar^{-1} are possible. Currently, phase changes of 10^{-4} radians can easily be detected by interferometers while an upper limit of 10^{-8} radians is most common for state-of-the-art devices. However, the advantage of high sensitivity can be offset by the sensitivity of systems to temperature and strain. Coherent sources (particularly laser diodes) suffer from intensity noise and must maintain adequate wavelength stability to minimise phase noise. The design of feedback circuits to drive the laser diodes and the difficulty in launching the light in monomode fibres are other practical difficulties encountered with set-ups using optical fibre interferometers. Because the interferometric output is sinusoidal in a phase detection system, for maximum sensitivity and linearity, a quadrature relationship is required between the measuring and reference phases. Furthermore, the output signal can be ambiguous unless there is constant referencing to a standard input condition.

Although the Michelson interferometer uses fewer components and is well suited for point measurements, severe problems can be encountered with source stability due to high levels of optical feedback which arise in such arrangements. This problem can be reduced in the Mach–Zehnder configuration, which also offers the advantage of having two antiphase outputs, a feature that is extremely useful in signal processing. Of the multiple beam configurations, the Fabry–Perot is the most common one [72]. Of the two-beam reciprocal-path interferometers, the most common one is the Sagnac, in which the change in symmetry between two counter-propagating beams is monitored so that angular velocity and the Faraday effect can be measured.

An example indicating the level of sophistication currently found in sensing applications employing interferometers is that of fibre optic gyros. In the simplest interferometric gyro, the phase difference $\Delta\phi$ between two counter-propagating light waves in a fibre coil of length L

and radius a at a rotation rate Ω is $\Delta\phi = 4\pi La\Omega/c\lambda$, where λ is the wavelength of the source used and c is the speed of light. In order to induce a phase difference between the two counter-propagating light waves to counteract the phase delay due to the fibre coil, so that feedback action can be applied to the system, a digital serrodyne (sawtooth with staircase slope) waveform may be used [73]. This is achieved by adjusting each step in the stair to be equal to one round-trip travelling time for the light wave in the fibre coil (Figure 3.22).

In order to reduce the length of fibres used without compromising the sensitivity, other fibre optic gyros based on high finesse ring resonators have been suggested. These utilise only a 10 m coiled optical fibre (preferably polarisation maintaining) and employ a highly coherent light source so that the ring resonator coupled to the rotating Sagnac interferometer produces Fabry–Perot like fringes. A variant of the ring resonator currently under development is the Brillouin fibre optic gyro which uses higher power in the fibre ring resonator to induce Brillouin scattering.

For intrinsic type sensors, phase modulation can also be achieved using coherent light in a multimode fibre. In the presence of strain in the fibre, the speckle pattern produced at the output changes due to multiple phase interactions and these pattern changes can be detected as an amplitude effect. Such schemes can be used for the dynamic measurements of vibrations, accelerations or acoustic fields.

Generally, for extrinsic sensor measurements, gas lasers are preferred because the lasing wavelength is accurately known, increasing the precision of the measurement, and also because higher output powers are available. However, recent advances in semiconductor laser diode output power are likely to increase their application in some extrinsic type sensors.

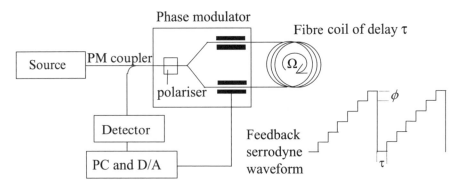

Figure 3.22   Interferometric fibre optic gyro using a digital serrodyne waveform to null the phase difference in the two counter-propagating waveforms, described in [73]

Of particular interest to accurate distance measurement when a large dynamic range and excellent resolution are required are the multiple and swept-wavelength techniques. These rely on the modulation of a carrier beam which is reflected from a target reflector. The phase of the returning modulation is measured relative to that of the outgoing reference signal. Progressively increasing the modulation frequency for each successive phase measurement results in an increase in the resolution of the measurement. In order to extend further the dynamic range of an interferometer, two sources with slightly different frequencies are used to produce the interferometric signature. This can be achieved either using two separate sources, or using a single source and the Zeeman effect, or using a Bragg cell in the reference path of the interferometer. In that case, the range is defined by the period of the visibility function, so that the dynamic range is increased by $\lambda_2/(\lambda_2 - \lambda_1)$ in comparison with that obtained with the interferometer illuminated by a single wavelength. Alternatively, by illuminating the interferometer sequentially by λ_1 and λ_2, or by using wavelength selective detectors, the difference in the two phases can be measured from $\phi(\lambda_1) - \phi(\lambda_2) = 2\pi n l(\lambda_2 - \lambda_1)/\lambda_1 \lambda_2$. It is theoretically possible to extend the range further using a greater number of wavelengths.

An alternative technique for the extension of range is the use of short coherence length sources. The technique makes use of the fact that when an interferometer is illuminated with a low-coherence source the interference can only be observed when the optical path differences are well balanced. Using a second interferometer in tandem with the first, and matching its path imbalance with the first, the position of zero path length imbalance may be uniquely determined. The interferogram in such a scheme comprises a group of interference fringes around the origin, corresponding to near-zero imbalance in the receiving interferometer, and two groups of smaller fringes that correspond to the receiving interferometer balancing the path length in the sensing interferometer.

The performance of well designed interferometric sensors is mainly limited by the shot and thermal noise in the photodetector. In systems where frequency noise and intensity noise are kept to a minimum, a phase resolution limit of approximately 10^{-3} rad $Hz^{-1/2}$ can be achieved. At low frequencies, $1/f$ noise and environmental noise degrade the sensitivity by several orders of magnitude. However, for extrinsic type sensors, the most significant contributor to degradation of fringe-contrast is the beam size variation with propagation. In sensor applications where fringe analysis is required, the analysis by Jin [74] and some of the recent series of proceedings of the Fringe Analysis Special Interest Group (FASIG) from the ISAT Group of the Institute of Physics (UK) may be used.

In environments where significant distortion of the waveform occurs, extrinsic type interferometric sensors become generally unsuitable. Occasionally, it may be possible, however, to fit the surface under observation with non-linear crystals capable of phase conjugation to eliminate the distortion (Figure 3.23).

For an incident wavefront $e^{i(\omega t + \phi)}$ on a non-linear crystal with a phase correction $e^{i2\omega t}$ the reflected waves will have a phase $e^{-i(\omega t + \phi)} e^{i2\omega t} = e^{i(2\omega t - \omega t - \phi)} = e^{i(\omega t - \phi)}$. The concept has already been demonstrated [75] at mm-wave frequencies using an electronic approach to conjugate the signal at specific antenna elements. Sampling at sufficiently dense intervals, it is possible to conjugate an entire wavefront.

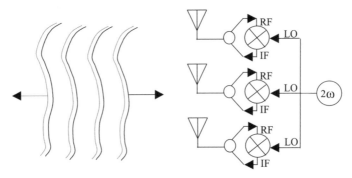

Figure 3.23   Phase conjugation concept demonstrated at mm-wave frequencies

Recently, it has also been realised that the application of force-feedback to the diaphragm in a conventional microphone leads to significant advantages in terms of linearity, frequency response and dynamic range. Feedback may be applied electrostatically by two electrodes on either side of the diaphragm [76]. However, this makes the normal technique of measuring the diaphragm displacement (dc excitation and change in capacitance) very difficult to apply. The use of optical interferometry avoids these problems. Although a Michelson configuration may be easier to implement [77], the Fabry–Perot arrangement has greater sensitivity, a more linear response and an in-line geometry allowing for the optical sensor to be inserted in a microphone housing (Figure 3.24).

For the application of force-feedback, if a voltage is applied across the plates of a parallel plate capacitor of area A, the resultant force would be proportional to the square of the applied voltage $F = A V^2 \varepsilon_0 / s^2$, where s is the electrode spacing and ε_0 is the permittivity constant. This, however, would lead to a non-linear response of the feedback loop. The application of a signal voltage V_s in the membrane and a polarisation voltage V_p to the electrodes (Figure 3.25), however, would result in a force given by $F = 2 A V_p V_s \varepsilon_0 / s^2$, which is proportional to the signal voltage.

Although simple op-amp circuits may be used to implement a feedback loop so that the diaphragm stays at the mid-point of a single fringe, careful design using discrete components and nested feedback loops can further extend the gain–bandwidth product, improving on the response time and the stability of the loop.

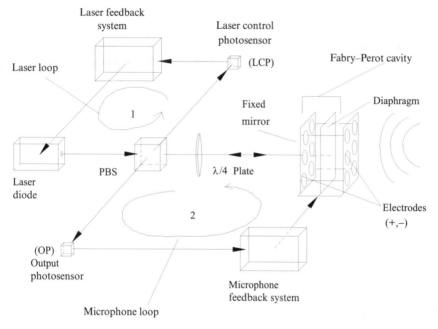

Figure 3.24 *Force-feedback optical microphone with intensity stabilisation of the laser diode (adapted from [77])*

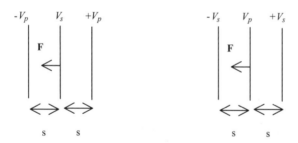

Figure 3.25 *Force-feedback using two electrodes to linearise the feedback loop and alternative arrangement of signals compensating for spacing differences*

Using electrostatic force-feedback it is possible to build a force-feedback microphone having the following characteristics: (i) a low tension, low mass diaphragm giving high responsivity at both low and high frequencies; (ii) a self-noise level close to the limit set by acoustic thermal noise, and a large dynamic range; (iii) precise control of the diaphragm using electrostatic force-feedback giving predictable frequency response and exceptional long term stability (dependent on the electrode spacing and the polarisation voltage only).

A higher corner frequency in the response of a microphone usually requires a higher tension of the diaphragm which leads to a lower sensitivity. The application of force-feedback, however, allows the use of a lower tension in the diaphragm, which increases the responsivity at lower frequencies. This also allows the use of a lighter diaphragm (as it has to be less strong), which results in higher responsivity at higher frequencies. The theoretical dynamic ranges of such an optical force-feedback driven microphone are 142 dB and 157 dB, assuming a 3 mW and a 100 mW laser diode power output, respectively (there is negligible phase noise in the system since the membrane is not moving), and compares favourably to 120 dB, which is a typical figure for a capacitive microphone. The lack of membrane motion implies that no dead-volume calibration is required when measurements of absolute pressures are performed and the overall distortion (higher order of Rayleigh modes of vibration on the membrane) is negligible. The predictable frequency response allows for direct spectral analysis of recorded acoustic emissions, of interest to non-destructive analysis studies.

3.5 Polarisation modulating sensors

The recent introduction of monomode fibres with special polarisation characteristics offers the possibility of producing a variety of sensors. Polarisation-maintaining, single-polarisation and ultra-low birefringent fibres are currently available on the market. The polarisation-maintaining fibre is a highly birefringent fibre, where the birefringence is the result of anisotropic thermal stress between the core and the stress-producing sectors in the fibre (bow-tie or panda types). The high birefringence exhibited by these fibres maintains the linearly polarised light, masking the effect of external bends or twists. In a similar manner, in single-polarisation fibres, one polarisation state suffers a high attenuation (50 dB/km) while the other suffers a low attenuation (5 dB/km). Single-polarisation fibres can be used as intrinsic sensors as the birefringence changes with both longitudinal strain and temperature. The state of polarisation can also be affected by the presence of electric or magnetic fields, and a variety of sensors can be realised using the Faraday magneto-optic effect, the electrogyration effect, the Kerr effect, the Pockels effect, the photo-elastic effect, or the optical activity of solutions.

During the last few years, evolution has been towards encapsulating the light paths in optical interferometers (Michelson, Mach–Zehnder, Sagnac), so as to act as compact stable elements that use phase rather than amplitude as the sensing parameter. An example of a null-balance Mach–Zehnder polarising interferometer implemented using monomode polarisation-maintaining fibres that may be implemented to measure the complex transmission coefficient of a sample is shown in Figure 3.26.

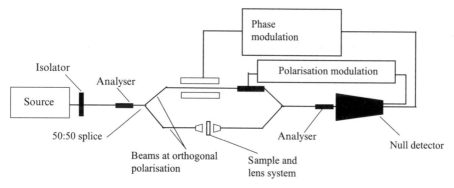

Figure 3.26   General diagram of a Mach–Zehnder null-balanced polarising interferometer implemented using monomode polarisation maintaining fibres

The bridge configuration is iteratively balanced by matching the phase and polarisation between the two arms (which have initially orthogonal polarisations) using feedback until a null is observed at the detector. Advantages of the nulling scheme suggested include immunity of the measurement to drifts in the source's intensity, as this would affect both arms equally, and the capability for absolute measurements using a detector that is not necessarily linear, as the amplitude transmission coefficient of the sample is directly related only to the state of polarisation in the reference arm.

Although telecommunications-grade fibres are used in most sensing applications, new fibre shapes have been developed maximising the advantages of high birefringence or polarisation maintenance. Particularly interesting configurations are the D-shaped fibres and the elliptically-shaped fibres. One of the advantages of the elliptical fibre is the simplicity of its construction compared to that of stress-induced birefringent fibres. By not using stress, the stress–temperature cross-sensitivity is reduced, increasing the stability of interferometric sensor applications. Furthermore, high order modes are azimuthally stable in elliptical core fibres and, therefore, over-moded fibre sensors become another possible application [78].

In most interferometric applications, when non-polarisation-maintaining fibres are used, the state of polarisation in the photodetector fluctuates due to thermal and mechanical perturbations in the fibre. These effects lead to undesirable intensity noise and deterioration of fringe visibility. As single-polarisation fibres are costly, polarisation controllers are a strong alternative. Since the state of polarisation has two degrees of freedom, at least two control elements are required for its implementation. If a precise (linear) polarisation is required at a particular location of the fibre, the controller may be followed by a polariser. In most of the polarisation controllers birefringence is induced by squeezing, bending or the Faraday effect. Most general controllers consist of two quarter-wave plates (each introducing a phase retardation difference between polarisation modes of 90 degrees) in series, and require active feedback.

Generally, fibre squeezers and Faraday rotators can provide only limited polarisation control and have either medium or fast response times but suffer from mechanical fatigue. Without active polarisation, it is possible to use a depolarisation scheme and apply polarisation filtering to the signal. Using metal-clad planar waveguides, a polarisation controller can be made because the light signal is coupled to the electric current it induces in the metal, which in turn suffers propagation losses via ohmic losses. Because the TM_0 mode has its electric field perpendicular to the metal surface, it is more strongly attenuated than the TE_0 mode. By directly coating the surface of a polished substrate surface with a conducting metal and by matching the group velocities of the optical and the electric (plasmon) surface waves, the performance of these devices can be further improved. Finally, as the metal-clad planar waveguides are difficult to fabricate, research efforts concentrate on the use of HB fibres in which one of the polarisation modes is more weakly guided than the other because of the existing stress-induced index anisotropy. D-shaped fibres and bent bow-tie fibres are also suitable for such applications.

A prerequisite for the proper analysis and understanding of polarisation-modulating sensors is familiarity with matrix algebra. A useful introduction to the use of matrix analysis in optical systems describing free space propagation and optical components in terms of Jones and Mueller matrices and the use of augmented matrices to account for errors in optical systems can be found in the book by Gerrard and Burch [79] and the use of the Poincaré sphere as a visualisation tool can be found in references [80] and [81].

A systematic analysis of monomode fibre optic sensors is presented in the article by Jones [82] and most recently a general treatment including couplers is presented in the article by Yu and Siddiqui [83]. Extension of these ideas to multimode sensors is possible provided that each propagating mode is treated as a separate 2×2 matrix so that an augmented matrix is formulated (an equivalent treatment exists for the description of beam modes in microwave antennas).

Recent advances in sources and detectors have allowed some of the techniques described above to be further extended to free-space propagating beams at the far-infra-red part of the spectrum where polarising grid beam splitters offer multi-octave operation with 99.9% efficiency, and the scattering due to surface irregularities is reduced. Such systems, however, very often require a train of lenses (made of high density polyethylene or TPX) to reduce the diffractive spreading of the beam as the size of the optical components of the system becomes similar to the wavelength. Null-balance bridge configurations measuring complex transmittance (a Martin–Puplett Mach–Zehnder hybrid polarising inter-ferometer), reflectance (a Martin–Puplett Michelson hybrid) and absorb-ance (using a combination of the two hybrids) can now be simply built on commercially available quasi-optical breadboards. These measure-ment techniques have much in common with ellipsometry (which is essentially a null-balancing technique) but have a larger throughput advantage because of the cylindrical symmetry of the instrument. Fur-thermore, they allow the simplest forms of the Fresnel equations to be used. Some further refinements to these techniques include the use of analyser grids and two detectors for each of the transmission and reflec-tion signals. This allows for the signatures to be separated into their co-polar and cross-polar components so that measurements of sample birefringence at mm and sub-mm wavelengths can be performed.

Finally, a major new trend in interferometry research is the construc-tion of monolithically integrated devices. Hofstetter *et al.* [84] produced a monolithically integrated displacement Michelson interferometer consisting of a DBR laser, a directional coupler, waveguides and a photo-detector, all fabricated on a GaAs/AlGaAs material. A measurement distance of 45 cm has been achieved using this configuration. Such advances offer considerable promise for future sensor miniaturisation and vibration immunity.

3.6 Distributed sensors

The field of distributed fibre optic sensing involves measuring the vari-ation of a physical or a chemical parameter along the length of the fibre. In many cases this technology is more attractive than using a large num-ber of separate point sensors (which provide information about the spa-tial variability of the measurand by means of an interpolation algorithm) to perform the same task and it also allows comparative measurements as a continuous function along the length of the fibre.

An important property of distributed fibre optic sensors is their ability to be multiplexed. Spatially multiplexed sensors are the simplest choice for many applications although the system requires many detectors, which in turn do not allow for inter-channel comparison, as the analogue signal can be affected by connector loss. Time-division multiplexing uses a single source and a detector but requires fast electronics or long optical path differences in the sensor cable. Wavelength-division multiplexing does not require high speed electronics but, again, multiple sources and detectors are necessary. In some applications, frequency-division multi-plexing of a sub-carrier has been used, showing a high resolution for a DC response.

Of the techniques currently available, optical time domain reflectom-etry (OTDR) is the most commonly employed technique for distributed sensing. This makes use of the finite propagation time of optical signals in the fibre and can therefore be considered as the optical waveguide analogue of the well-known radar technique. Of the OTDR techniques available at the moment, the Rayleigh backscattering technique is the most common and uses commercially available OTDR equipment. The Raman technique is based on measuring the Stokes/anti-Stokes ratio and although it gives a much weaker signal than Rayleigh scattering the signal is independent of fibre type or composition, and has very low sensitivity to undesirable parameters. A main disadvantage of Raman OTDR is the slow time response due to the averaging technique used and the requirement of sophisticated electronics which translate into a large cost for the base unit. Fluorescent OTDR, optical frequency domain reflectometry (OFDR) and optical amplification by counter-propagating pulses are other alternative techniques currently requiring further development.

Two-way propagation methods have also been mentioned in the literature, whereby a fibre supporting two guided modes of different velocity is used, and mode cross-coupling is monitored in response to an external influence. Using a frequency-swept source the beat frequency of the cross-coupled signals between these modes can be detected at the far end of the fibre, providing information regarding the position of cross-coupling. Furthermore, using two-way propagation in a fibre-loop and performing a phase comparison of the light emerging from each counter-propagating path, it is possible to get information about the position at which the path length disturbance occurs and the rate of change of the disturbance.

In certain applications, a disadvantage of distributed fibre optic sensing is the limited signal-to-noise ratio compared to point sensors, because of the limited amount of energy available at a particular location for the transduction principle to occur. Commercially available OTDR units based on the Rayleigh signal have a dynamic range of 20 dB and a resolution of 0.01 dB (Figure 3.27). Because of the trade-off between signal-to-noise ratio and spatial resolution, it may be preferable to have quasi-distributed systems where the transduction takes place only in certain locations along a fibre, and to process the information using a single or multiple detector system.

Alternatively, correlation techniques (Figure 3.28) may be implemented to improve on the spatial resolution of a distributed or quasi-distributed sensor. A correlation technique tackles the problem of limited source power by improving the mean launched power, but not the peak power as in other techniques.

In order to perform correlation, the laser diode is intensity modulated by a pseudo-random bit sequence $S_1(t)$ having the duration $T_0 = (2^n - 1)T_b$, where T_b is the bit length. The backscattered signal is detected by a photodiode, amplified and multiplied by a time shifted sequence

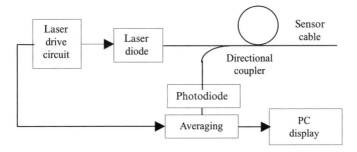

Figure 3.27   OTDR standard signal averaging technique

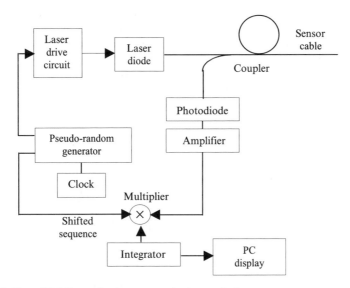

Figure 3.28   OTDR employing a correlation technique

$S_2(t - \theta)$. The product signal is integrated during a time $T = kT_0$, a multiple of T_0, and the cross-correlation of the two sequences is

$$R(\theta) = \int_0^T S_1(\theta) \cdot S_2(t - \theta)\mathrm{d}t \qquad (7)$$

The spatial resolution is determined by the width (FWHM) of the cross-correlation peak, corresponding to the bit duration of the sequence. More recently, Geiger and Dakin [85] have shown an alternative technique of interrogating the range of reflective markers in an optical fibre, further enhancing the signal-to-noise ratio and improving spatial resolution. As pointed out by Culshaw [86], for point sensor network topology applications, the measured area can either be determined by the application of the sampling theorem (the engineering approach) or by the application of Green's functions (the physics approach).

Basic network topologies include linear arrays, ring, reflective star, reflective tree, transmissive star and ladder configurations with the maximum number of sensors required, and the robustness of the application, determining the network topology. The interrogation principle (time, frequency or wavelength division multiplexing), the input power, the noise sources, the fibre optic principle, the sensor dynamic range and the power per sensor required determine the maximum number of multiplexable sensors. Recently, Connelly [87] showed that current constraints regarding multiplexing may be relaxed in the future using an optical preamplification technique.

Time division multiplexed sensors operate by launching short pulses of light into a transmissive (ladder) or reflective (linear array or star) network, and detecting the returning pulses that are delayed by a time interval dependent on the sensor length, the fibre refractive index and the velocity of light in free space. In a typical fibre ($n = 1.5$), light travels a distance of 1 m in approximately 5 ns, and therefore care must be taken so that the pulse duration and pulse period are not larger than the echo-return time from the most remote sensor or connector.

Alternatively to time gating, the propagation delay can be converted into a beat frequency by applying a frequency modulation (usually a ramp) to the optical source. Using a beam-splitter, part of the emitted light and the reflected light from the sensor are then combined and detected using an electrical spectrum analyser that separates the beat frequency of each sensor according to its time delay. The main advantage of optical frequency domain reflectometry (OFDR) over OTDR is the greater average optical channel power and narrower allowable receiver bandwidth, both of which result in a better SNR and hence higher sensitivity and spatial resolution. Furthermore, the continuous wave operation of the laser results in much lower source noise compared to the pulsed operation, and avoids the detector saturation often encountered in OTDR systems.

An alternative technique particularly used with ladder sensor networks is the frequency division multiplexing (FDM) scheme. With FDM, every sensor in the network is assigned a frequency channel, within which the sensor signal may be modulated in amplitude, frequency or phase by the corresponding measurand. Using sub-carrier frequency division multiplexing, a network of N intensity-encoded reflective or transmissive sensors can be modulated at N different frequencies, and the sum signal is detected. The amplitude and phase of each sensor are then plotted in a phasor diagram, the vector angle being related to the phase shift (time delay) between each sensor, and the length of each vector being proportional to the attenuated signal (determined using a lock-in amplifier). In order to determine the unknown amplitudes, the system is operated at all N frequencies simultaneously, and the measured components form a set of N linearly independent equations, whose components are the N amplitudes and phases. The technique described, although complicated, is particularly useful for distributed control applications where the matrix of the system under control is equivalent to the matrix equivalent of the sensor.

When frequency encoded multiplexed sensors (such as microresonators) are used, optical excitation (continuous wave or pulsed) can be provided with a swept frequency waveform. Normally, the individual sensor frequencies (detected with an RF spectrum analyser) can vary only within a small range of frequencies in response to the measurand. As a result, by careful resonator design, the sweep frequency does not overlap with the sensor's dynamic range and allows dead bands, to avoid crosstalk.

Wavelength division multiplexing (WDM) is another technique which is similar to FDM, its major advantage being that it utilises the large bandwidth of the fibre optic signal. Similarly to spectral encoding, the sensors can be multiplexed in the time or frequency domain. The technique requires a set of lasers or LEDs operating at different wavelengths to be combined into a single input fibre, and then transmitted through a wavelength dispersive component, that will distribute the optical energy according to wavelength to the different sensors. The modulated intensities from the sensors are normally recombined using a demultiplexer and a detector array or a scanning monochromator. From a power budget point of view, a large number of sensors may be addressed with WDM, and crosstalk is the limiting factor. Crosstalk in WDM is generally due to either overlap of component (filter grating and coupler) characteristics or overlap in the sensor signal channels.

Finally, alternative spread spectrum techniques based on the correlation technique described earlier but allowing for many sensors to be multiplexed [88], as well as hybrid TDM/WDM [89] and coherence multiplexing techniques [90, 91], have also been suggested.

From an applications point of view, current developments in quasi-distributed sensors are concentrating mainly on the Bragg grating sensor technology, and an overview of the most commonly used techniques can be found in the article by Kersey *et al.* [92]. Recent advances in the fabrication of quality Bragg reflectors with different grating conditions (linear, chirped, etc.), signal processing using hybrid interferometer stabilisation schemes to measure the shift in wavelength [93] and new Fabry–Perot filters made from very thin porous silicon layers deposited over emitting microcavities are likely further to enhance the possible range of sensor applications.

Furthermore, recent advances in the field such as the geometric representation of errors in measurements of strain and temperature [94, 95] and the papers by Brady *et al.* [61] and Davis and Kersey [96] suggest that the stress component can be separated by the temperature component along the length of a fibre. Such results should further increase the potential uses of quasi-distributed and distributed sensors in the future. In applications where a birefringent single-mode fibre is available, a spectral polarimetric technique may also be used [97].

An alternative choice to the Bragg grating sensor is the microbend sensor which is based on light attenuation due to sinusoidal distortions of fixed periodicity (a resonance condition) in a multimode optical fibre. These sensors can provide either totally distributed or quasi-distributed measurements. Based on the microbending properties of fibres, a variety of sensors has been described in the literature for measuring acceleration [98], strain [99, 100], pressure [101], displacement [102, 103], acoustic waves [104] and pH [105]. A historical review of fibre optic microbend sensors can be found in the article by Berthold [106]. Although a theoretical analysis of the modal dispersion induced by stresses in a multimode plastic optical fibre has been recently reported by Zubia and Arrue [107], these fibres have an unsuitable modulus of elasticity for most sensing applications. Finally, a recent interesting variation is the spiral fibre microbend sensor [108], which is a hybrid between a microbend and a macrobend sensor.

Advances in OTDR resolution and the rapid advancement of new fibre-coating techniques with hydrogels offer the possibility of producing a distributed chemical sensor for detecting the presence of water. More recently [109], the range of possible applications of such a hydrogel based sensor has been extended to measure water potentials of soils and salt solutions. The sensor is based on measurements of the changes in the Rayleigh backscattering signal induced by quasi-sinusoidal distortions of fixed periodicity on a multimode graded index optical fibre. In the presence of water molecules, the swelling action of a poly(ethylene oxide) based hydrogel coating, which is deposited on a support cable, causes the fibre to deform locally by squeezing it against a kevlar thread, helically wound along the length of the sensor cable (Figure 3.29).

The resonant period for the microbend is equal to the pitch length of the kevlar thread and is given from $\Lambda_m = 2\pi r/\sqrt{(2\delta)}$, where r is the core radius and δ is the maximum refractive index difference between the core and cladding, and the system operates at a resonant mode (only odd harmonics produce attenuation).

It can be assumed that the observed attenuation caused by the microbending of the optical fibre is due to the work done by the hydrogel, as it expands by absorbing water. For a hydrogel piece immersed in a solution, under constant temperature and pressure, the minimum amount of work needed to cause a change from one state to another is called the Gibbs free energy, ΔG, of the system. Furthermore, to every chemical component in a solution, we can assign a free energy per mole (J mol^{-1}) of that species. This quantity is called the chemical potential, μ_j, of species j. The chemical potential depends on the randomness (entropy) of the system, and concentration is a quantitative description

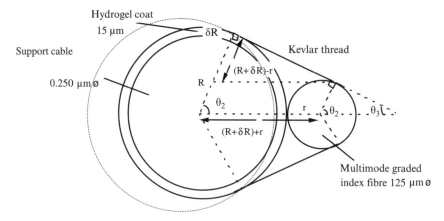

Figure 3.29 *Geometric model of support cable, hydrogel in dry (solid line) and swollen wet (dotted line) state and multimode graded index fibre*

of this randomness. The net diffusion of species j proceeds in the direction of decreasing μ_j, which is the same as that of decreasing concentration. Pressure differences (hydrostatic pressure) and the presence of charged particles (electrical potential) can also affect this net diffusion process. Therefore, the chemical potential of any species j may be represented by the following sum:

$$\mu_j = \mu_j^* + RT\ln a_j + \bar{V}_j P + z_j FE + m_j gh \tag{8}$$

where μ_j^* is a reference level, R is the gas constant, and T is temperature (at 25°C RT is 2.479×10^{-3} m^3 MPa mol^{-1}). The activity term a_j (which may be regarded as corrected concentrations) is due to molecular interactions of the species j with themselves and is related to the concentration of the solutes c_j by means of an activity coefficient $a_j = \gamma_j c_j$. The $\bar{V}_j P$ term represents the effect of pressure (in excess of atmospheric) on chemical potential. Finally, $\bar{V}_j$ is the partial molal volume of species j, z_j is the charge number of species j, F is the Faraday constant, E is the electrical potential and $m_j gh$ is a gravitational term. Generally, the presence of solutes in an aqueous solution tends to decrease the activity of water, and leads to an osmotic potential ψ in that solution, given in Reference [110], $\psi = -kRT\sigma\phi$, where ϕ is the osmotic coefficient (determined from isopiestic measurements), k is the number of species (e.g. for NaCl $k = 2$) and σ is the moles of solute per 1000 g of solvent.

There is a direct relation between concentration, water potential, chemical potential and therefore Gibbs energy in a solution, and under equilibrium conditions, the Gibbs energy of the solution should be in equilibrium with the Gibbs energy in the hydrogel. The total change in free energy ΔG in the hydrogel results from mixing a pure penetrant (swelling agent) and an amorphous isotropic polymer network. The swelling properties of the polymer are controlled by the combination of free energies of mixing, ΔG_{mix}, between the solvent and the polymer chains, and also by the elastic response (elastic free energy), ΔG_{el}, of the network to volume increases due to solvent absorption

$$\Delta G = \Delta G_{mix} + \Delta G_{el} \qquad (9)$$

Under constant temperature, ΔG_{el} is a function of the effective number of chains in the network and an expansion factor, which expresses the linear deformation of the network structure due to isotropic swelling. Furthermore, ΔG_{mix} is a function of the polymer–solvent volume fractions and the Flory polymer–solvent interaction parameter. The observed change in attenuation in the sensor is a function dependent on the resonant mode condition, the geometry of the sensor and the degree of swelling of the hydrogel $P(t) = f(\Delta G_{el})$, and therefore, knowledge of ΔG_{mix} for a particular change in ΔG of a solution can be used directly to link ΔG_{el} to the osmotic potential of the solution.

In an aqueous environment, the deformer (hydrogel) responds to a change in water potential $\Delta \psi$ by applying a force to the fibre causing the amplitude of the fibre deformation X to change by an amount ΔX. The transmission coefficient for light propagating through the bent fibre I is in turn changed by an amount ΔI so that

$$\Delta I = \left(\frac{\Delta I}{\Delta X}\right) D \Delta \psi \qquad (10)$$

where D is given by $D \Delta \psi = \Delta X$ and ΔX is related to the amount of displacement of the fibre core inside the hydrogel. The coefficient $\Delta I / \Delta X$ relates the change in transmission to the change in fibre deformation amplitude and expresses the sensitivity of the optical fibre to microbending losses. Critical parameters in the $\Delta I / \Delta X$ term are the core and cladding refractive indices and their dependence on temperature, the type of fibre buffer coating used and the optical power distribution between modes in the fibre. Within the operational range of the system developed at Strathclyde, it may be assumed that $\Delta I / \Delta X = 12$ dB km^{-1} μm^{-1}. For dc measurements, a change in $\Delta \psi$ is related to the photodetector signal output by

$$i_s = \frac{qe W_0}{h\omega}\left(\frac{\Delta I}{\Delta X}\right) D \Delta \psi \qquad (11)$$

where h is Planck's constant, ω is the optical frequency, q is the detector quantum efficiency, e is the electron charge, and W_0 is the input optical power. For a shot-noise-limited detection system, the mean square photodetector noise is given by

$$i_n^2 = 2e \frac{qe W_0 I}{h\omega} \Delta f \qquad (12)$$

where Δf is the post-detection bandwidth. The smallest osmotic pressure that can be detected is

$$\Delta \psi_{min} = D^{-1} \left(\frac{\Delta I}{\Delta X} \right)^{-1} \left(\frac{2Ih\omega \Delta f}{qW_0} \right)^{1/2} \tag{13}$$

The distributed nature of the sensor makes it particularly useful for a wide range of applications, such as to monitor salinity and water quality in wells, for 'surrogate' measurements of water potentials and possibly for measuring soil water potentials.

Finally, similarly to the already described hydrogel coated fibre, other custom made twin-core fibres, liquid-core fibres, rare-earth metal doped fibres and special coating fibres are currently under investigation by companies and research groups. Although these fibres can have very good performance as point or small distance (a few metres) quasi-distributed sensors, their suitability as truly distributed sensors is limited because of the current difficulties in maintaining uniform properties over large fibre lengths during the manufacturing process.

3.7 Further reading

BORN, M., and WOLF, E.: 'Principles of optics' (Pergamon Press, Oxford, 1980)

CULSHAW B.: 'Optical fibre sensing and signal processing' (Peter Peregrinus, 1984)

CULSHAW, B., and DAKIN, J. (Eds): 'Optical fiber sensors', vols. I and II, 1988; vols. III and IV, 1997 (Artech House)

DAKIN, J. P. (Ed.): 'The distributed fibre optic sensing handbook' (IFS Publications, UK, Springer-Verlag, 1990)

SNYDER, A. W., and LOVE, J. D.: 'Optical waveguide theory' (Chapman & Hall, London, 1983)

UDD, E. (Ed.): 'Fiber optic sensors, an introduction for engineers and scientists' (Wiley Interscience Publications, 1991)

USHER, M. J., and KEATING, D. A.: 'Sensors and transducers, characteristics, applications, instrumentation, interfacing' (Macmillan Press Ltd., 1996, 2nd edn.)

UTTAMCHANDANI, D., and ANDONOVIC, I. (Eds): 'Principles of modern optical systems', vol. II (Artech House, 1992)

3.8 References

1 CONLEY, M. P., ZAROBILA, C. J., and FREAL, J. B.: 'Reflection type fiber-optic sensor', SPIE conference on *Fiber-optic and laser sensors IV*, 1986, **718**, pp. 237–243

2   HOCHBERG, R. C.: 'Fiber-optic sensors', *IEEE Transactions on Instrumentation and Measurement*, 1986, **35**, pp. 447–450
3   KROHN, D. A.: 'Intensity modulated fiber optic sensors: an overview', SPIE conference on *Fiber-optic and laser sensors IV*, 1986, **718**, pp. 2–11
4   TOBA, E., SHIMOSAKA, T., and SHIMAZU, H.: 'Non-contact measurement of microscopic displacement and vibration by means of fiber optics bundle', SPIE conference on *Fiber-optic and laser sensors IX*, 1991, **1584**, pp. 353–363
5   HU, A., and CUOMO, F. W.: 'Theoretical and experimental study of a fiber optic microphone', *Journal of the Acoustical Society of America*, 1992, **91**, pp. 3049–3057
6   FUJII, H., ASACURA, T., MATSUMOTO, T., and OHURA, T.: 'Output power distribution of a large core optical fiber', *Journal of Lightwave Technology*, 1984, **2**, pp. 1057–1062
7   HE, G., and CUOMO, F. W.: 'Displacement response, detection limit and dynamic range of fibre-optic lever sensors', *Journal of Lightwave Technology*, Fiber-Optic Metrology and Standards, 1991a, **9**, pp. 1618–1625
8   HE, G., and CUOMO, F. W.: 'A light intensity function suitable for multimode fiber-optic sensors', *Journal of Lightwave Technology*, 1991b, **9**, pp. 545–551
9   COOK, R. O., and HAMM, C. W.: 'Fiber optic lever displacement transducer', *Applied Optics*, 1979, **21**, pp. 3230–3241
10 SCHNEITER, J. L., and SHERIDAN, T. B.: 'An optical tactile sensor for manipulators', *Robotics and Computer-Integrated Manufacturing*, 1984, **1**, pp. 65–71
11 POWELL, A. J.: 'A simple two-fiber optical displacement sensor', *Review of Scientific Instruments*, 1974, **45**, pp. 302–303
12 CONFORTI, G., BRENCI, M., MENCAGLIA, A., and MIGNANI, A. G.: 'Fiber-optic thermometric probe utilizing GRIN lenses', *Applied Optics*, 1989, **28**, pp. 577–580
13 BRENCI, M., MENCAGLIA, A., and MIGNANI, A. G.: 'Fiber-optic sensors for simultaneous and independent measurement of vibration and temperature in electric generators', *Applied Optics*, 1991, **30**, pp. 2947–2951
14 CUSWORTH, S. D., and SENIOR, J. M.: 'A reflective optical sensing technique employing a GRIN rod lens', *Journal of Physics E: Scientific Instruments*, 1987, **20**, pp. 102–103
15 GYS, T., and NORDVIK, J. P.: 'A simple noncontact multimode fiberoptic proximity sensor', *SPIE*, 1985, **586**, pp. 197–202
16 GREENHOW, R. C.: 'Simple low-cost optical position transducer', *Journal of Physics E: Scientific Instruments*, 1987, **20**, pp. 602–604
17 SPILLMAN, W. B., JR., and McMAHON, D. H.: 'Frustrated-total-internal-reflection multimode fiber-optic hydrophone', *Applied Optics*, 1980, **19**, pp. 113–117
18 SPILLMAN, W. B., and GRAVEL, R. L.: 'Moving fiber-optic hydrophone', *Optics Letters*, 1980, **5**, pp. 30–31
19 MURZATA, G., and SENIOR, J. M.: 'Referenced intensity-based optical fibre sensors', *International Journal of Optoelectronics*, 1994, **9**, pp. 339–349
20 CULSHAW, B., FOLEY, J., and GILES, I. P.: A balancing technique for

optical fiber intensity modulated transducers', in KERSTEN, R. T. and KIST, R. (Eds), OFS 84, second international conference on *Optical fiber sensors*, 1984, pp. 117–120

21 WANG, H., HENGSTERMANN, T., REUTER, R., and WILLKOMM, R.: 'Robust improvement in the signal-to-noise of laser fluorosensor through selectively averaging signals', *Review of Scientific Instruments*, 1992, **63**, pp. 1877–1879

22 LAMMERINK, T. S. J., and FLUITMAN, J. H. J.: 'Measuring method for optical fibre sensors', *Journal of Physics E: Scientific Instruments*, 1984, **17**, pp. 1127–1129

23 SPILLMAN, W. B., and LORD, J. R.: 'Self-referencing multiplexing technique for fiber-optic intensity sensors', *Journal of Lightwave Technology*, 1987, **5**, pp. 865–869

24 COCKSHOTT, C. P., and PACAUD, S. J.: 'Compensation of an optical fibre reflective sensor', *Sensors and Actuators*, 1989, **17**, pp. 167–171

25 LIBO, Y., and ANPING, Q.: 'Fiber-optic diaphragm pressure sensor with automatic intensity compensation', *Sensors and Actuators A*, 1991, **28**, pp. 29–33

26 KALYMNIOS, D.: 'Linear and scalable optical sensing technique for displacement', in GRATTAN, K. T. V. and AUGOUSTI, A. T. (Eds): 'Sensors IV: Technology systems and applications' (Institute of Physics Publishing, Bristol and Philadelphia, 1993), pp. 291–294

27 WOBSCHALL, D., and HEJAZI, S.: 'Aperture intensity ratio fiber-optic displacement sensors', *Review of Scientific Instruments*, 1987, **58**, pp. 1543–1544

28 BORSBOOM, P. C. F., and TEN BOSCH, J. J.: 'Fiber-optic scattering monitor for use with bulk opaque material', *Applied Optics*, 1982, **21**, pp. 3531–3535

29 JONES, B. E., and SPOONCER, R. C.: 'Two-wavelength referencing of an optical fibre intensity-modulated sensor', *Journal of Physics E: Scientific Instruments*, Rapid Communication, 1983, **16**, pp. 1124–1126

30 SPILLMAN, W. B.: 'Industrial uses of fiber optic sensors', SPIE Conference on *Fiber-optic and laser sensors IV*, 1986, **718**, pp. 21–27

31 MARTIN, L. R., and HILL, M. W.: 'Optical measurement of aqueous kinetics at micromolar concentrations' *Journal of Physics E: Scientific Instruments*, 1987, **20**, pp. 1383–1387

32 MURUGAIAH, V., and SYNOVEC, R. E.: 'Molecular weight sensing of polyethylene glycols by flow injection analysis and refractive index gradient detection', *Analytica Chimica Acta*, 1991, **246**, pp. 241–249

33 CHIRON, B.: 'Highly efficient plastic optical fluorescent fibers and sensors', SPIE conference on *Plastic Optical Fibers*, 1991a, **1592**, pp. 86–95

34 LAGUESSE, M., and BOURDINAUD, M.: 'Characterisation of fluorescent optical fibers for X-ray beam detection', SPIE conference on *Plastic optical fibers*, 1991, **1592**, pp. 96–107

35 CHIRON, B.: 'Anamorphosor for scintillating plastic optical fiber applications', SPIE conference on *Plastic optical fibers*, 1991b, **1592**, pp. 158–164

36 WOLFBEIS, O.: 'Chemical sensing using indicator dyes', in DAKIN, J. and CULSHAW, B. (Eds): 'Optical fiber sensors vol. IV,' (Artech House, 1997)

37 STEWART, G., and CULSHAW, B.: 'Optical waveguide modelling and

design for evanescent field chemical sensors', *Optical and Quantum Electonics*, 1994, **26**, pp. S249–S259

38   STEWART, G., JIN, W., and CULSHAW, B.: 'Prospects for fibre optic evanescent-field gas sensors using absorption in the near-infrared', *Sensors and Actuators B – Chemical*, 1997, **38**, pp. 42–47

39   STEWART, G., MUHAMMAD, F. A., and CULSHAW, B.: 'Sensitivity improvement for evanescent wave gas sensors', *Sensors and Actuators B*, 1993, **11**, pp. 521–524

40   STEWART, G., and JOHNSTONE, W.: 'Evanescently coupled components', in CULSHAW, B. and DAKIN, J. (Eds): 'Optical fiber sensors, vol. III' (Artech House, 1997)

41   McCULLOCH, S., and UTTAMCHANDANI, D.: 'Development of a fibre optic micro-optrode for intracellular pH measurements', *IEE Proceedings – Optoelectronics*, 1997, **144**, pp. 156–161

42   CAHN, T. M.: 'Biosensors', Sensor Physics and Technology Series (Chapman & Hall, 1993)

43   BADINI, G. E., GRATTAN, K. T. V., TSEUNG, A. C. C., and PALMER, A.W.: 'Porous glass substrates with potential applications to fiber optic chemical sensors', *Sensors and Materials*, 1998, **10**, pp. 29–46

44   DRESS, P., BELZ, M., KLEIN, K. F., GRATTAN, K. T. V., and FRANKE, H.: 'Physical analysis of teflon coated capillary waveguides', *Sensors and Actuators B – Chemical*, 1998, **51**, pp. 278–284

45   BADINI, G. E., GRATTAN, K. T. V., and TSEUNG, A. C. C.: 'Impregnation of a pH sensitive dye into sol-gels for fiber optic chemical sensors', *Analyst*, 1995, **120**, pp. 1025–1028

46   FOWLES, M., and WAYNE, R. P.: 'Ozone monitor using an LED source', *Journal of Physics E: Scientific Instruments*, 1981, **15**, pp. 1143–1145

47   MITCHELL, D. N., and WAYNE, R. P.: 'A photometer for measuring small changes in optical absorption', *Journal of Physics E: Scientific Instruments*, 1980, **13**, pp. 494–495

48   DRESS, P., BELZ, M., KLEIN, K. F., GRATTAN, K. T. V., and FRANKE, H.: 'Water-core waveguide for pollution measurements in the deep ultraviolet', *Applied Optics*, 1998, **37**, pp. 4991–4997

49   MacCRAITH, B., GRATTAN, K. T. V., CONNOLLY, D., BRIGGS, R., BOYLE, W. J. O., and AVIS, M.: 'Cross-comparison of techniques for monitoring of total organic carbon (TOC) in water sources and supplies', *Water Science and Technology*, 1993, **28**, pp. 457–463

50   BENJATHAPANUN, N., BOYLE, W. J. O., and GRATTAN, K. T. V.: 'Binary encoded 2nd differential spectrometry using UV-Vis. Spectral data and neural networks in the estimation of species type and concentration', *IEE Proceeding – Science Measurement and Technology*, 1997, **144**, pp. 73–80

51   BOISDE, G. E., and HARMER, A.: 'Chemical and biochemical sensing with optical fibers and waveguides' (Artech House, 1997)

52   HADJILOUCAS, S., KARATZAS, L. S., KEATING, D. A., and USHER, M. J.: 'Humidity measurement using a plastic optical fibre sensor', in GRATTAN, K. T. V., and AUGOUSTI, A. T. (Eds): 'Sensors VII, technology systems and applications' (IOP Publications, 1995) pp. 230–235

53   AUGOUSTI, A. T., GRATTAN, K. T. V., and PALMER, A. W.: 'Visible

LED pumped fiber-optic temperature sensor', *IEEE Transactions on Instrumentation and Measurement*, 1988, **37**, pp. 470–472

54 AUGOUSTI, A. T., GRATTAN, K. T. V., and PALMER, A. W.: 'A liquid-crystal fibre-optic temperature switch', *Journal of Physics E: Scientific Instruments*, 1988, **21**, pp. 817–819

55 KAJANTO, I., and FRIBERG, A. T.: 'A silicon-based fibre-optic temperature sensor', *Journal of Physics E: Scientific Instruments*, 1988, **21**, pp. 652–656

56 KRUGER, A.: 'Light-emitting diodes as dynamic pressure transducers', *Journal of Physics E: Scientific Instruments*, 1985, **18**, pp. 944–946

57 ZHANG, Z. Y., GRATTAN, K. T. V., PALMER, A. W., and MEGGITT, B. T.: 'Thulioum doped intrinsic fiber-optic sensor for high temperature measurements(>1100 degrees C)', *Review of Scientific Instruments*, 1998, **69**, pp. 3210–3214

58 COLLINS, S. F., BAXTER, G. W., WADE, S. A., SUN, T., GRATTAN, K. T. V., ZHANG, Z. Y., and PALMER, A. W.: 'Comparison of fluorescence-based temperature sensor schemes: Theoretical analysis and experimental validation', *Journal of Applied Physics*, 1998, **84**, pp. 4649–4654

59 LYNCH, B.: 'Photoacoustic resonators', in 'Fibre optic sensors', IOP Short Meeting Series No 7, 1987, Institute of Physics

60 CULSHAW, B.: 'Optical excited resonant sensors', in 'Fibre optic sensors', IOP Short Meeting Series No 7, 1987, Institute of Physics

61 BRADY, G. P., KALLI, K., JACKSON, D. A., REEKIE, L. and ARCHAMBAULT, J. L.: 'Simultaneous measurement of strain and temperature using the first- and second-order diffraction wavelengths of Bragg gratings', *IEE Proceedings – Optoelectronics*, 1997, **144**, pp. 156–161

62 HONG, J., HUANG, W.-P., and MAKINO, T.: 'Analytical theory of coupled waveguide reflectors and resonant transmission filters', *IEE Proceedings – Optoelectronics*, 1995, **142**, pp. 209–218

63 LOPETEGI, T., FALCONE, F., MARTINEZ, B., GONZALO, R., and SOROLLA, M.: 'Improved 2–D optimized photonic bandgap microstrip'. *23rd Int. Conf. Infrared and millimeter waves*, 1998, PARKER, T. J., and SMITH, S. R. P. (Eds), pp. 197–198

64 BRAND, G. F.: 'Millimeter-wave beam control by the illumination of a semiconductor'. *23rd Int. Conf. Infrared and millimeter waves*, 1998, PARKER, T. J., and SMITH, S. R. P. (Eds), pp. 248–249

65 BROOKS, P., and REEVE, C. D.: 'Limitations in acousto-optic FM demodulators', *IEE Proceedings – Optoelectronics*, 1995, **142**, pp. 149–156

66 KIM, B. Y., BLAKE, J. N., ENGAN, H. E., and SHAW, H. J.: 'All-fiber acousto-optic frequency shifter', *Optics Letters*, 1986, **11**, pp. 389–391

67 JONES, J. D. C., and BARTON, J. S.: 'Fiber-optic sensors for condition monitoring and engineering diagnostics', in CULSHAW, B. and DAKIN, J. (Eds): 'Optical fiber sensors, vol. III' (Artech House, 1997), pp. 207–259

68 LOCKEY, R. A., and TATAM, R. P.: 'Multicomponent time division multiplexed optical fibre laser Doppler anemometry, demodulators', *IEE Proceedings – Optoelectronics*, 1997, **144**, pp. 168–175

69 JACKSON, D. A., PRIEST, R., DANDRIDGE, A., and TVETEN, A. B.: 'Elimination of drift in a single mode optical fiber interferometer using a piezoelectric stretched coiled fiber', *Applied Optics*, 1980, **19**, pp. 2926–2929

70 JARZYNSKI, J.: 'Frequency response of a single-mode optical fiber phase modulator utilizing a piezoelectric plastic jacket', *Journal of Applied Physics*, 1984, **55**, pp. 3243–3250

71 CULSHAW, B.: 'Optical fiber transducers and applications', Sensors 82, *Transducer technology and temperature measurement*, vol. 3, Sensor developments and applications, pp. 1–14

72 VAUGHAN, J. M.: 'The Fabry–Perot interferometer' (Adam Hilger, Bristol and Philadelphia, 1989)

73 HOTATE, K.: 'Fiber-optic gyros', in CULSHAW, B. and DAKIN, J. (Eds): 'Optical fiber sensors, vol. III' (Artech House, 1997), pp. 167–206

74 JIN, W., UTTAMCHANDANI, D., and CULSHAW, B.: 'Dynamic phase measurement over a large dynamic range in a fibre optic Sagnac interferometer', *International Journal of Optoelectronics*, 1993, **8**, pp. 57–65

75 CHANG, Y., FETTERMAN, H. R., NEWBERG, H. R., and PANARETOS, S.K.: 'Millimeter phase conjugation using artificial nonlinear surfaces', *Applied Physics Letters*, 1998, **72**, pp. 745–747

76 KEATING, D. A.: 'Force feedback microphone', *Proceedings of the Institute of Acoustics.*, 1986, **8**, (6), pp. 67–73

77 KARATZAS, L. S., KEATING, D. A., and USHER, M. J.: 'A practical optical force-feedback microphone', *Transactions of the Institute of Measurement and Control*, 1994, **16**, pp. 75–85

78 DYOTT, R. B.: 'Elliptical fiber waveguides' (Artech House, 1995)

79 GERRARD, A., and BURCH, J. M.: 'Introduction to matrix methods in optics' (Dover Publications, 1994)

80 JERRARD, H. G.: 'Transmission of light through birefringent and optically active media: the Poincaré sphere', *Journal of the Optical Society of America*, 1954, **44**, pp. 634–640

81 CULSHAW, B.: 'Optical fibre sensing and signal processing' (Peter Peregrinus, 1984)

82 JONES, J. D. C.: 'Monomode fibre optic sensors', WILLIAMS, D. C. (Ed.): 'Optical methods in engineering metrology' (Chapman & Hall, 1993), pp. 415–464

83 YU, A., and SIDDIQUI, A. S.: 'Systematic method for the analysis of optical fibre circuits', *IEE Proceedings – Optoelectronics*, 1995, **142**, pp. 165–175

84 HOFSTETTER, D., ZAPPE, H. P., and DANDLIKER, R.: 'Optical displacement measurement with GaAs/AlGaAs-based monolithically integrated Michelson interferometers', *Journal of Lightwave Technology*, 1997, **15**, pp. 663–670

85 GEIGER, H., and DAKIN, J. P.: 'Low-cost high resolution time-domain reflectometry for monitoring the range of reflective points', *Journal of Lightwave Technology*, 1995, **13**, pp. 1282–1288

86 CULSHAW, B.: 'Smart structures and materials' (Artech House, 1995)

87 CONNELLY, M. J.: 'Optically preamplified interferometric fibre optic acoustic sensor multiplex characteristics', in AUGOUSTI, A.T. (Ed.): 'Applied optics and optoelectronics', (IOP Publications, 1998) pp. 135–140

88 EVERARD, J. K. A.: 'Novel signal processing techniques for enhanced OTDR sensors', SPIE conference on *Fiber-optic sensors*, The Hague, 1987, **798**, p. 42

89 KERSEY, A. D., and DANDRIDGE, A.: 'Demonstration of a hybrid time/wavelength division multiplexed interferometric fiber sensor array', *Electronics Letters*, 1991, **27**, p. 554

90 BROOKS, J. L., WENTWORTH, R. H., YOUNGQUIST, R. C., TUR, M., KIM, B. Y., and SHAW, H. J: 'Coherence multiplexing of fiber-optic interferometric sensors', *Journal of Lightwave Technology*, 1985, **3**, pp. 1062–1072

91 KERSEY, A. D., and DANDRIDGE, A.: 'Phase noise reduction in coherence multiplexed interferometric fiber sensors', *Electronics Letters*, 1986, **22**, pp. 616–618

92 KERSEY, A. D.: 'Multiplexing techniques for fiber-optic sensors', in CULSHAW, B., and DAKIN, J. (Eds): 'Optical fiber sensors, vol. III' (Artech House, 1997), pp. 369–407

93 SHI, W. J., NING, Y. N., GRATTAN, K. T. V., and PALMER, A. W.: 'Novel hybrid interferometer stabilization scheme used in wavelength shift measurement for Bragg grating sensors', *Review of Scientific Instruments*, 1998, **69**, 1961–1965

94 JIN, W., MICHIE, W. C., THURSBY, G., KONSTANTAKI, M., and CULSHAW, B.: 'Simultaneous measurement of strain and temperature: Error analysis', *Optical Engineering*, 1997a, **36**, pp. 598–609

95 JIN, W., MICHIE, W. C., THURSBY, G., KONSTANTAKI, M., and CULSHAW, B.: 'Geometric representation of errors in measurements of strain and temperature', *Optical Engineering*, 1997b, **36**, pp. 2272–2278

96 BRADY, G. P., KALLI, K., JACKSON, D. A., REEKIE, L., and ARCHAMBAULT, J. L.: 'Simultaneous measurement of strain and temperature using the first- and second-order diffraction wavelengths of Bragg gratings', *IEE Proceedings – Optoelectronics*, 1997, **144**, pp. 156–161

97 TANAKA, S., and OHTSUKA, Y.: 'Fibre optic spectral polarimetry for sensing multiple stress-loaded locations along a length of fibre', *IEE Proceedings – Optoelectronics*, 1997, **144**, pp. 156–161

98 FREAL, J. B., ZAROBILA, C. J., and DAVIS, C. M.: 'A microbend horizontal accelerometer for borehole deployment', *Journal of Lightwave Technology*, 1987, **5**, 993–996

99 ASAWA, C. K., YAO, S. K., MOTA, N. L., and DOWNS, J. W.: 'High sensitivity fibre-optic strain sensors for measuring structural distortion', *Electronics Letters*, 1982, **18**, pp. 362–364

100 WEISS, J. D.: 'Fiber-optic strain gauge', *Journal of Lightwave Technology*, 1989, **7**, pp. 1308–1318

101 BERTHOLD, III, J. W., GHERING, W. L., and VARSHNEYA, D.: 'Design and characterisation of a high temperature fiber-optic pressure transducer', *Journal of Lightwave Technology*, 1987, **7**, pp. 870–876

102 LAGAKOS, N., LITOVITZ, T., MACEDO, P., MOHR, R., and MEISTER, R.: 'Multimode optical fibre displacement sensor', *Applied Optics*, 1981, **20**, pp. 167–168

103 LAGAKOS, N., COLE, J. H., and BUCARO, J. A.: 'Microbend fibre-optic sensor', *Applied Optics*, 1987, **26**, pp. 2171–2180

104 FIELDS, J. N., and COLE, J. H.: 'Fibre microbend acoustic sensor', *Applied Optics*, 1980, **19**, pp. 3265–3267

105 MICHIE, W. C., CULSHAW, B., KONSTANTAKI, M., McKENZIE, I., KELLY, S., GRAHAM, N. B., and MORAN, C.: 'Distributed pH and water detection using fibre-optic sensors and hydrogels', *Journal of Lightwave Technology*, 1995, **13**, pp. 1415–1420

106 BERTHOLD, III, J. W.: 'Historical review of microbend fiber-optic sensors', *Journal of Lightwave Technology*, 1995, **13**, pp. 1193–1199

107 ZUBIA, J., and ARRUE, J.: 'Theoretical analysis of the modal dispersion induced by stresses in a multimode optical fibre', *IEE Proceedings – Optoelectronics*, 1997, **144**, pp. 397–403

108 YOSHINO, T., INOUE, K. and KOBAYASHI, Y.: 'Spiral fibre microbend sensors', *IEE Proceedings – Optoelectronics*, 1997, **144**, pp. 145–150

109 HADJILOUCAS, S., MICHIE, W. C., CULSHAW, B., KONSTANTAKI, M., KEATING, D. A., USHER, M. J., GRAHAM, N. B., and MORAN, C.R.: 'Hydrogel based distributed fiber-optic sensor for measuring soil salinity and soil water potentials'. Paper no. 9, IEE Colloquium on *Progress in fibre optic sensors and their applications*, ref. no.: 1995/194, November 1995

110 LANG, A. R. G: 'Osmotic coefficients and water potentials of sodium chloride solutions from 0 to 40°C', *Australian Journal of Chemistry*, 1967, **20**, p. 2017

Chapter 4

Amorphous semiconductor photoreceptors and X-ray image sensors

S. M. Vaezi-Nejad

4.1 Introduction

Amorphous materials can best be defined as materials in which the three dimensional periodicity is absent. The atomic arrangement is not entirely random as in gases and there is a short range order of a few Ångström units. As with crystalline solids, amorphous materials can be insulators, semiconductors or metals, and in some cases at very low temperatures they can even be superconductors [1, 2].

One of the outstanding optical applications of amorphous semiconductors is in electrophotography, commonly known as xerography [3–5]. The photosensitive insulating device employed in xerography is known as a photoreceptor. As schematically illustrated in Figure 4.1, there are five major steps involved in the xerography process: (a) sensitisation of the photoreceptor by a corona discharge; (b) image-wise exposure of the photoreceptor to produce a latent electrostatic image; (c) development of the latent image by fine toner particles; (d) transfer of the developed image to plain paper; and (e) fixation of the transferred image by fusing [6].

Improved methods of transferring print to paper have been established for applications such as word processing, electronic mail and high quality facsimile. Laser printers satisfy the desirable characteristics of high speed, graphic flexibility and high print quality because of the electronic controls and laser optics involved. He–Ne lasers together with external acousto-optic modulators may be used to modulate the light beam with the information to be printed.

The exposure unit for laser printers is more sophisticated than that of a conventional photocopier. The optical system focuses and scans a laser beam in a raster-like fashion across the moving photoreceptor. When

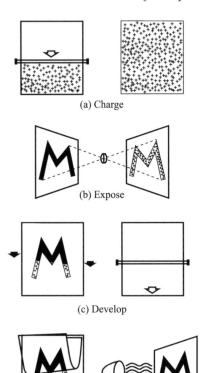

(a) Charge

(b) Expose

(c) Develop

(d) Transfer   (e) Fix

Figure 4.1   The basic xerographic process

solid state light sources such as GaAlAs lasers are used instead of a gas laser, the complete exposure unit in a laser printer would be replaced with a unit consisting mainly of three sub-systems: a laser diode, an electronic circuit to drive the laser diode and to stabilise the light output, and an optical system to collimate the divergent light beam [7]. Laser diodes offer advantages of high reliability, low cost, small size, low power consumption, low voltage supply and direct modulation to several 100 MHz.

The aim of this chapter is to describe the design, manufacture, properties and application of amorphous semiconductors as image sensors. After introducing the relevant concepts and background theories in Section 4.2, the preparation, structure, properties and performance of two important classes of amorphous semiconductors will be described in some detail. These are amorphous selenium based (a-Se) and hydrogenated amorphous silicon (a-Si : H) photoreceptors. In the next section X-ray imaging techniques and application of amorphous semiconductors as X-ray image sensors will be discussed. Recent advances in optical application of amorphous semiconductors are briefly explained in the final section.

4.2 Relevant concepts and background theories

There are a number of desirable characteristics that a useful photoreceptor should exhibit:

(i) Good charge acceptance. A photoreceptor of thickness ~50 μm should be capable of sustaining a surface voltage of ~500 V, which corresponds to an electric field of ~10^5 V cm^{-1}.
(ii) Slow rate of charge decay in the dark. The photoreceptor must be a good insulator in the dark to retain the charge on its surface long enough for the completion of xerographic image transfer.
(iii) Rapid discharge when exposed to light.
(iv) Low residual voltage. The rapid photodischarge in (iii) terminates in a residual voltage. As will be explained in detail later, in order to achieve a good xerographic image, the residual voltage must be very small.

The above functions should be performed by the photoreceptor under repetitive cycles; otherwise, the photoreceptor is said to be fatigued. Most fatigue effects generally disappear after a rest period.

Finally, there are other desirable characteristics such as photosensitivity to a particular wavelength, crystallisation resistance, temperature and humidity resistance, thickness uniformity and a mechanically hard surface.

4.2.1 Charge acceptance

The electrical resistivity of photoreceptors in the dark must be very high so that they can be charged by a corona device to a high initial potential. Thus a practical photoreceptor can be treated as a parallel plate capacitor with the photoconductive insulating layer as the dielectric medium. The surface charge density Q_s is given by

$$Q_s = \varepsilon_0 \varepsilon_r V_0^2 / L \qquad (1)$$

where ε_0 is the permittivity of free space, ε_r is the dielectric constant of the insulating layer, V_0 is the applied voltage across the photoreceptor and L is the thickness of the insulating layer. Equation 1 shows that for a given thickness, a higher surface potential requires a greater amount of surface charge. Therefore if an equal amount of charge is deposited on two photoreceptors of the same material but different thicknesses, the potential across the thicker photoreceptor will be higher. For formation of an electrostatic image, it is necessary to apply sufficient surface charge to establish a high electric field $E_0 = V_0/L$, which is normally 10^5 V cm^{-1}.

Corona devices utilising pin or thin wires as charge emitters are commonly used for depositing the surface charge. The quality of the final image depends in part upon the operational limitations of the corona device. Under-charging of the photoreceptor will not allow utilisation of maximum sensitivity and contrast, while over-charging can produce fatigue effects and surface defects. The effectiveness of sensitisation depends on three main factors [8–10]:

(i)   the magnitude of the current produced by the corona;
(ii) the amount of current which reaches the photoreceptor surface;
(iii) the uniformity with which the current is deposited on to the photoreceptor surface.

A well designed corona will produce a pre-selected current, evenly distribute it over the deposited area of the plate and maintain this consistently for a long period of time. In an attempt to find the most efficient device for the xerographic time-of-flight experiment which is described later, the author investigated various device configurations utilising a sharp pin or pins and wire or wires as the corona emitter.

Three corona devices are shown in Figure 4.2. The charge distribution of the devices was studied using a current sensing probe. As illustrated in Figure 4.3 the best current distribution was obtained for a positive applied voltage to a double wire corona emitter. Note that d is the distance between the corona emitter and the detecting probe. These curves clearly show that devices based on thin wires (~50 μm thick) provide a more uniform charge distribution [9, 10]. Table 4.1 shows the charge acceptance of a number of amorphous semiconductor photoreceptors [11]. It should be noted that the doping concentration in Table 4.1 and throughout the whole chapter is expressed as wt%.

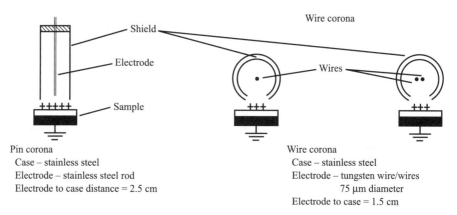

Figure 4.2   Schematic sketches of different corona devices

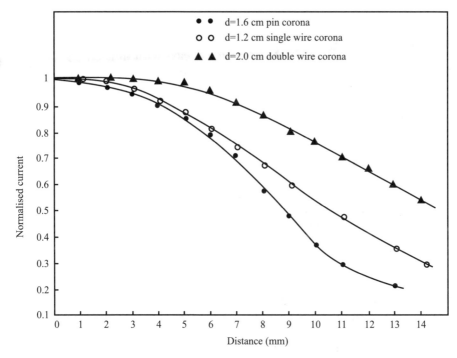

Figure 4.3   Normalised current distribution for applied voltage of +7 kV

Table 4.1   Charge acceptance of various a-Se based photoreceptors

Content	Thickness single-layer (μm)	Thickness double-layer (μm)		Charge acceptance (V)
		PGL	CTL	
Se + 3.5% Te	68	–	–	1010
Se + 5% Te	64	–	–	890
Se + 7% Te	62	–	–	810
Se + 10% Te	63	–	–	760
Se + 15% Te	61	–	–	700
Se + 2.3% Te + 20 ppm Cl	60	–	–	1100
Se + 3.5% Te + 16 ppm Cl	58	–	–	930
Se + 4.6% Te + 26 ppm Cl	60	–	–	980
Se + 5% Te + 16 ppm Cl	62	–	–	950
Se + 12.5% Te + 20 ppm Cl	66	–	–	900
Se + 4.5% Te/Se	–	11.5	46.5	1230
Se + 6.7% Te/Se	–	10	48	1105
Se + 10% Te/Se	–	6	51	1030
Se + 15% Te/Se	–	7	49	890
Se + 17% Te/Se	–	5	51	840

4.2.2 Dark decay

During the sensitisation, a photoreceptor with its substrate grounded is charged to a surface potential $V(t)$ which decays with time and is referred to as the dark decay. V_0 in eqn. 1 is the initial surface potential $V(t)$ when $t = 0$, and the maximum possible value of V_0 for a photoreceptor is known as the charge acceptance. There are several factors which influence the dark decay. The most significant are [12–14]:

(a) carriers released thermally within the insulating layer (in the case of a single or a mono-layer structure) or layers (in the case of a multi-layer structure);
(b) carriers injected at the free surface of the photoreceptor, referred to as the surface injection current;
(c) carriers injected at the interface of the photoreceptor insulating layer with the metal substrate.

a-Se based multi-layer photoreceptors, which will be described in detail in Section 4.3, generally consist of a transport layer, a thin photo-generation layer and sometimes a third layer of pure or very lightly doped a-Se as the voltage sustaining layer (commonly called the protection layer). In these photoreceptors, the contribution from source (a) is different when compared with the single-layer structures. Here, the bulk generation current is the total contribution of all the various layers constituting the photoreceptor. However, contributions from the other two sources (b) and (c) are similar to that of a single layer provided that they are of the same impurity content and fabricated under similar conditions. It should be evident from these considerations that the mechanism of dark decay in single-layer photoreceptors, particularly a-Se, is an important ingredient in understanding the dark decay in a-Se based multi-layer photoreceptors.

By introducing a blocking layer between the metal substrate and the photoconductive insulating layer, it is possible to limit the sources of the dark discharge to a surface charge injection and bulk charge generation. In single-layer photoreceptors based on amorphous semiconductors, as the transit time of the free charge carriers is very small compared to the time scale of the dark decay process [15], the analysis of dark decay can be approximated to a basic capacitor problem. A dark current I flows through the photoconductive insulating layer as the dielectric and causes the dark decay. The rate of change of the surface potential $V(t)$ is thus given by

$$\mathrm{d}V(t)/\mathrm{d}t = I/C \tag{2}$$

where C is the capacitance of the photoreceptor. For an electric field $E(t) = V(t)/L$ (V cm^{-1}), surface generation rate J_s (cm^{-2} s^{-1}) and bulk generation g_B (cm^{-3} s^{-1}), the above equation becomes

$$dE/dt = q(J_s + g_B L)/\varepsilon_0\varepsilon_r \qquad (3a)$$

For bulk generation of only one type of carrier, the bulk current is reduced by half and eqn. 3a becomes

$$dE/dt = q(J_s + \tfrac{1}{2} g_B L)/\varepsilon_0\varepsilon_r \qquad (3b)$$

Since the measurable parameter dE/dt is a linear function of thickness L and the proportionality constant is the bulk generation term, it is possible to evaluate independently the surface and the bulk contributions [8, 12].

4.2.3 Photo-induced discharge

After the sensitisation, the charged photoreceptor is exposed to light reflected from a document to create an electrostatic image. In the case of single-layer structures, in the illuminated areas, the photoreceptor potential is discharged due to the photocurrent that flows perpendicular to the surface. For positive surface potential, which is normally the case, photogenerated electrons will neutralise the positive surface charge while free holes drift across the photoconductor layer to neutralise the induced negative charge on the aluminium substrate. The lateral conductivity of the photoreceptor is negligible, ensuring the sharpness of the latent image. When an exposing light beam falls on the photoreceptor, the surface potential decays towards a low value which depends on the exposure energy per unit area.

In multi-layer photoreceptors, charge carrier generation and its subsequent transport are carried out in separate layers [16]. In the popular double-layer structures, the charge generation layer (CGL) can form the top or bottom layer. If the CGL is the top layer, the visible light is absorbed close to the floating surface of the photoreceptor. On the other hand, if CGL forms the bottom layer (e.g. see Figure 4.8), the charge transport layer (CTL) is transparent to allow the light to pass through essentially unattenuated. However, in both cases some of the absorbed light energy is used for the formation of electron–hole pairs in the photo-generation layer (PGL) and the rest is converted into heat. Photons constituting the light each carry energy hc/λ, where h is Planck's constant, c is the speed of light in a vacuum and λ is the wavelength of light.

The apparatus required for measuring the photo-induced discharge normally consists of the following components:

  (i)   a light source operating at a known colour temperature, for example a tungsten lamp at 3200 K;
 (ii)   a monochromator or a filter for selecting the desired wavelength λ;
(iii)   a photocell for measuring the light intensity;

(iv) a transparent voltage probe to measure the photo-induced surface
      voltage.

If desirable, the spectral response curve can also be obtained from the
plot of discharge rate against wavelength.

4.2.4 Carrier range and Schubweg

The carrier range R is defined as the product of the carrier mean free
time τ and the drift mobility μ. Thus the hole and electron ranges are
given by $\mu_p \tau_p$ and $\mu_n \tau_n$, respectively. Both μ and τ should be large enough
for the carriers not to suffer significant trapping during their transit
across the photoreceptor. Determination of these parameters is
described in detail in Section 4.3.3. If the carrier range is not sufficiently
large, the photoreceptor will not produce a good image. Practical values
measured by the author for the carrier range in some common photoreceptors are shown in Table 4.2. A more common parameter deduced
from the carrier range is known as the Schubweg s, which is the product
of the carrier range and electric field, i.e.

$$s = \mu \tau E \qquad (4)$$

This is the mean distance travelled by a charge carrier before it is
trapped. If s is comparable to or less than the photoreceptor thickness,
then a considerable number of the injected carriers will be trapped,
causing unacceptably high residual voltage and degradation of the
image produced. Values of hole Schubweg s_h (expressed in terms of the
sample thickness L) in several a-Se based photoreceptors are also shown
in Table 4.2. Note that in all cases $s_h \gg L$, indicating negligible carrier
trapping.

Table 4.2   Carrier range and hole Schubweg in a-Se based photoreceptors

Photoreceptor	Thickness L (μm)	Carrier range (10^{-5} cm^2/V)	Hole Schubweg as multiple of L
a-Se	58	17	29 L
a-Se + 3.5% Te	68	10	15 L
a-Se + 5% Te	64	11	17 L
a-Se + 3.7% Te + 10 ppm Cl	58	22	37 L
a-Se + 4.6% Te + 26 ppm Cl	62	31	53 L

4.2.5 Residual voltage

As mentioned earlier, the resistivity ρ of the photoreceptor should be
sufficiently large to achieve a high charge acceptance and long dielectric

relaxation time τ_r. It is essential for τ_r, given by $\varepsilon_0 \varepsilon_r \rho$, to exceed the image development time. However, high ρ is not the only requirement. When s becomes less than the sample thickness, the photoreceptor cannot usually be completely discharged and a residual potential occurs. Often, the residual voltage cannot be erased and increases with cycling. As mentioned earlier, s should therefore be large in comparison to the photoreceptor thickness L. Consider a well-known single-layer structure such as a pure a-Se photoreceptor. In the sensitisation exposure part of the xerographic cycle, the photoreceptor is charged by a corona and then discharged by exposure to strongly absorbed light. Depending on charging polarity (usually positive), one sign of the carrier acts to neutralise charge on the top surface while the opposite sign drifts through the bulk towards the ground substrate. As a result of this carrier displacement, the surface voltage decays in time – in other words, it is photo-induced discharged. The surface voltage would decay to zero were it not for the fact that some fraction of the carriers in transit through the bulk is captured by deep traps, resulting in a small voltage called the residual voltage V_r. Space charge neutrality is eventually re-established over a much larger time scale. Using a practical apparatus consisting of components similar to those described in Section 4.3.3, it is possible to obtain information about these deep traps [12, 17].

From the double integration of Gauss's law, the residual voltage can be related to the space charge distribution function $\rho_s(x')$ by a one dimensional equation [11, 18]

$$V_r = (\varepsilon_0 \, \varepsilon_r)^{-1} \int dx \int \rho_s(x') dx' \tag{5}$$

where ε_r is the dielectric constant.

For a single homogeneous sample without dielectric discontinuity, the space charge distribution in the bulk ρ_s is uniform so that $\rho_s(x') = \rho_0$ and V_r is given by

$$V_r = \rho_0 L^2 / 2\varepsilon_0 \, \varepsilon_r \tag{6}$$

where L is the photoreceptor thickness.

If the total density of deep traps is N_t, then, when the residual voltage saturates under repeated charging and discharging cycles, $\rho_0 = q N_t$, i.e. all deep traps are filled. Equation 6 becomes

$$V_{r\infty} = q N_t L^2 / 2\varepsilon_0 \, \varepsilon_r \tag{7}$$

Other models have also been proposed to describe the residual voltage build-up [19, 20].

4.2.6 Practical considerations

A basic experimental apparatus developed by the author for measuring dark decay, photo-induced discharge, residual voltage and saturated

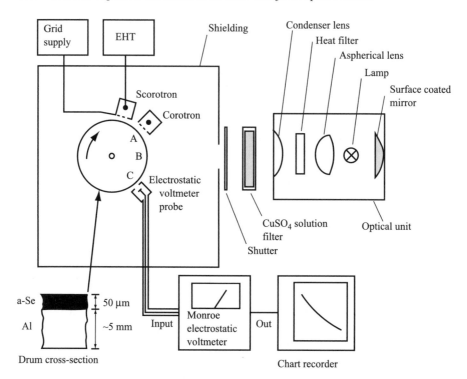

*Figure 4.4   A practical experimental apparatus for dark decay and residual
voltage measurements*

residual voltage is shown in Figure 4.4. The photoreceptor under test is
mounted on a turntable driven by an ac motor. The floating surface of
the photoreceptor is charged by a corotron at station A. For a high initial
voltage (several hundred volts) the output of an HT supply is connected
to the corotron. For a lower initial voltage (a few hundred volts or less),
the scorotron is used. In this latter case, a potential is applied to a grid
placed between the corona charger and the floating surface of the photoreceptor to ensure that the initial voltage on the photoreceptor does not
exceed the required level. The photoreceptor is uniformly charged by
rotating it several times under the corona discharge, after which the latter
is switched off. The surface potential is monitored on an electrostatic
voltmeter. The photoreceptor, the corona charging unit and the voltmeter probe are housed in a well shielded and dark environment. The
output of the electrostatic voltmeter is connected to a chart recorder
which continuously records the surface voltage against time. For saturated residual voltage measurements, the charged sample is discharged by
a strongly absorbed short wavelength light through a shutter. The discharge lamp shown is a 250 W, 24 V quartz halogen lamp mounted in a
projector lamp assembly. A heat filter is incorporated to remove most of

the IR radiation to avoid heating the sample surface. Note that $CuSO_4$ solution is used as a blue-pass filter. Thus the photoreceptor is charged at position A, discharged at position B by an intense light and its residual voltage is measured at position C. This cycle is continuously repeated until the residual voltage saturates.

As illustrated in Figure 4.5, during the continuous repetition of the xerographic cycle, the residual voltage builds up and after many cycles it saturates.

For a-Se single-layer photoreceptors with thicknesses varying from 41 to 58 μm, the positive saturated residual voltage has been measured to be $+48$ to $+74$ V [18]. Using eqn. 7, these measurements indicate the presence of $1–2 \times 10^{13}$ cm^{-3} deep traps. Similar measurements on a-Se:Te single-layer photoreceptors with different wt% concentrations of Te (4.5%, 6.7%, 10%, 15% and 17%) have revealed the effect of Te on the density of deep traps. Variation of the saturated positive residual voltage $V_{r\infty}$ with the Te concentration and the corresponding changes in the density of deep traps N_t are shown in Figure 4.6.

Note that the saturated residual voltage initially increases with Te, and beyond 6–8 wt% Te it begins to fall. The density of deep hole traps calculated from eqn. 5 above exhibits a similar behaviour. There are two plausible explanations: (i) because of monatomic decreases in the band-gap with Te composition, the release time from deep traps may not

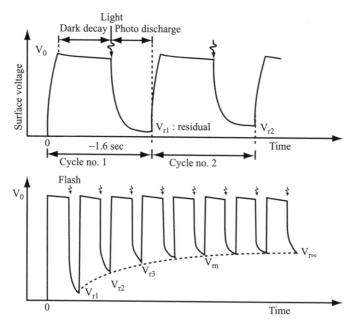

Figure 4.5   *Build-up of surface potential for two successive cycles (top curve) and under repeated cycles (lower curve)*

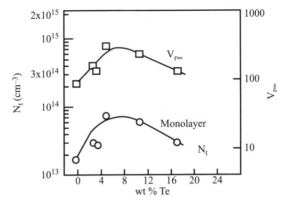

Figure 4.6   Positive saturated residual voltage and the density of deep traps against Te concentration in a-Se : Te photoreceptors

remain sufficiently long in comparison to the charging and discharging cycle time to allow an eventual saturation of the deep traps – as Te is added, the release time becomes shorter and a small fraction of N_t gets filled, hence resulting in the fall in $V_{r\infty}$; (ii) the rapid bulk thermal generation of holes and their sweep-out during the time between charging and discharging result in a volume density of negative ions. Recombination of a photo-injected hole with a negative ion will convert the latter to a neutral site and hence will not result in a detectable potential. It should be noted that in multilayer structures, eqn. 7 is no longer valid because of the dielectric mismatch and possible interface trapping [18–20].

4.3 Photoreceptors

There are a variety of photoreceptors in use, mainly due to the increase in the number of companies producing photocopiers and non-impact printers. Photoreceptors are of different substrates, flexibility, photo-sensitivity, speed and, of course, cost. In general photoreceptors may be divided into three broad categories of (i) amorphous semiconductors, (ii) organic and (iii) semiconductor power-resin binder layers. Since the first category is the main subject of this chapter, the other two categories are briefly described first.

4.3.1 Organic photoreceptors

There is a considerable interest in organic photoreceptors, mainly because unlike amorphous semiconductors they can be readily prepared in a flexible configuration [6, 21–24]. Common organic photoreceptors are either single-layer or double-layer configurations. Figure 4.7 shows

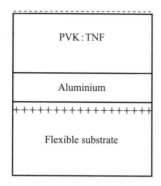

Figure 4.7   A PVK:TNF single-layer organic photoreceptor

one of the early single-layer structures which was based on a charge transfer complex of polyvinyl carbazole (PVK) as the hole transport and 2,4,7-tri-nitro-9-fluorenone (TNF) as the electron transport [21]. The photoconductive layer of PVK:TNF is ~10 μm thick and is prepared on a grounded aluminium sheet and a flexible substrate. When the charge surface of the photoreceptor is exposed to radiation of appropriate wavelength (He–Ne laser), the absorbed light creates electron–hole pairs which, under the action of an applied high electric field, are separated. The holes migrate via PVK molecules to the negatively charged floating surface of the photoreceptor, while the electrons migrate via TNF to the aluminium ground. It is possible to add photogenerating dye such as chlorodiane blue to the PVK:TNF in a dispersed manner to increase further the sensitivity of the photoreceptor. The dye absorbs the incident light, creating additional extra electron–hole pairs [6, 14].

Separation of charge generation and charge transport functions in a double-layer photoreceptor allows additional optimisation of the photoconductive layer. In double-layer organic photoreceptors, the CGL (charge generation layer) is usually an organic dye or pigment whereas the CTLs (charge transport layers) are comprised of donor or acceptor molecules in a polymer [3, 14]. CGLs are typically between 0.5 and 5 μm in thickness and coated adjacent to the substrate. The charge transport layers are usually coated over the generation layer and are between 10 and 30 μm in thickness. Figure 4.8 shows a typical double-layer structure in which CTL is a 10 μm thick pytazoline based material transparent to visible light and a 0.1 μm thick CGL of chlorodiane blue. The light passes unattenuated through the CTL to reach the CGL where it creates electron–hole pairs [23].

In organic photoreceptors, the mobility of charge carriers is much smaller than in amorphous semiconductor photoreceptors. In fact, the carrier transit time in organic photoreceptors can be so large that it is comparable to the processing times of practical interest, thus limiting the

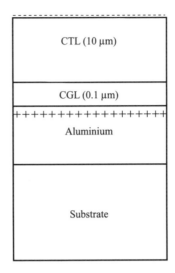

Figure 4.8   A double-layer organic photoreceptor

use of such photoreceptors. In order to reduce the transit time, the applied electric field is increased. For example, single-layer organic photoreceptors are typically ~15 μm thick and usually charged to ~500 V, giving a high internal electric field. However, such a high field can make the photoreceptor susceptible to localised breakdown. When holes and electrons are mobile, cyclic instabilities due to the build-up of a space charge are minimised. Single-layer organic photoreceptors should therefore be bipolar so that they can be used with either a positive or negative surface potential having comparable levels of sensitivity. Organic photoreceptors can be made uniform and of a relatively low dielectric constant offering advantages of easy sensitisation. A lower charge density can be used on the photoreceptor surface which in turn requires a lower exposure.

There is a variety of new organic materials used as xerographic photoreceptors [3, 14, 24]. In most of these materials, holes or electrons (but not both) are mobile. As only one carrier is mobile, penetrating or weakly absorbed exposures create free and deeply trapped carriers of opposite polarity, thus creating volume or bulk space charges. Change in the electrical properties with cycling is normally attributed to volume space charges. A solution to this problem is the well-known technique of separating the functions of the photoreceptor into two separate layers of CTL and CGL. Table 4.3 gives a summary of various materials for organic photoreceptors [3, 14, 24]. Note that the materials for CTL are divided into two categories according to their suitability as hole transport materials or electron transport materials. Similarly, for the CGL, materials are divided into the two categories of molecular complex and pigments.

Table 4.3 Summary of various materials for organic photoreceptors

Comments	Materials for the CTL		Materials for the PGL	
	Hole transport	Electron transport*	Molecular complex†	Pigments
PVK stands for poly(N-vinylcarbazole)	Compound class: arylakanes, arylamines, hydrazones, oxadiazoles	Alkyl-substituted nitrated fluorene-9-ones	PVK & TNF	Azos (chloro-diane blue,
			Dye–polymer aggregate	Tautomer, Trisazo, Perinone bisazo)
TNF stands for 2,4,7-trinitro-9-fluorenone		Polyester molecule (with Ti Opc as PGL)		Aromatic diamines
				Phthalocyanines
* Development of electron transport has been very slow and there are no commercial photoreceptors based on this system.		Polymers doped with mixtures of donor and acceptor molecules	–	Squaraines
				Polycyclic aromatics
† Molecular complexes are limited.				

For practical applications in instruments the physical properties of organic photoreceptors are not as useful as their mechanical properties. For example, they have poor scratch and thermal resistance. It is for these reasons that such photoreceptors in particular are used in laser printers and low volume copiers where long life times and very fast response times are not major considerations. These photoreceptors, particularly the single layers, have low cost potential in both processing and the necessary equipment. Other advantages are high radiation sensitivity and very low thermal generation rates, the latter permitting the development of infra-red-sensitive materials. As mentioned earlier, these photoreceptors can be made into rolls, sheets, sleeves or belts and can even be welded.

4.3.2 Semiconductor powder–resin binder layers

Semiconductor powder–resin binder layers are two-phase photoreceptor materials consisting of fine semiconductor powder bound together mechanically by insulating resins such as acryl, silicone, polyurethane and polyvinyl acetate [25–27]. Zinc oxide (ZnO) and cadmium sulphide (CdS) are the most common binder type photoreceptors which are limited to low volume/low speed copiers due to fatigue problems. ZnO is a direct band-gap II–VI semiconductor with a band-gap of 3.2 eV at room temperature. The photoresponse of ZnO can be extended to wavelengths longer than the fundamental absorption edge by the absorption of organic dyes. There is no single model for dye sensitisation applicable to all the ZnO–dye systems. For example, in vacuum-evaporated thin films (~0.1 μm) of ZnO and various dyes, it has been found that, under dark conditions, the electrons are transferred from ZnO to the ZnO–dye interface forming a Schottky barrier [28]. Upon illumination some of these trapped electrons are injected back into ZnO. Evaporation of a merocyanine dye on to ZnO single crystals (prism face) under an ultra-high vacuum results in an increase in the dark surface conductivity. Studies on the sensitisation of ZnO single crystals by Rose Bengal, erythrosin B and fluorescein dyes show that there is no charge transfer between the ZnO and the dyes in air under dark conditions [25, 26]. Since the sensitisation was greatly reduced in a vacuum, the investigators concluded that the presence of oxygen was essential [25, 26].

Cadmium sulphide is also a direct band-gap II–VI semiconductor with a band-gap of 2.4 eV at room temperature. The room temperature values of electron mobility in undoped CdS crystals of high resistivity have been reported to lie between 0.026 and 0.029 $m^2V^{-1}s^{-1}$. The values of hole mobility at room temperature have been measured to be within a smaller range of 0.001–0.002 $m^2V^{-1}s^{-1}$. From the analysis of transient photocurrent, the electron mobility in single crystals of CdS is estimated to be greater than 0.006 $m^2V^{-1}s^{-1}$ and the coresponding lifetime has been calculated to be less than 0.9 μs. Undoped CdS normally possesses a low photosensitivity. An increase in the photosensitivity has been achieved by compensation of the donors with acceptor impurities. For example, the spectral response has been found to extend to longer wavelengths by doping with copper. It has been reported that the electronic processes involved in negative corona charging of CdS powder–resin binder layers are accompanied by significant and reversible chemical reactions of the surface of the photoreceptor [27]. The electron range of CdS-resin binder layers has been reported to lie between 3.3 and 5.6×10^{-7} cm^2 V^{-1}. The overall conclusion of a detailed systematic investigation on ZnO and CdS powder–resin layers [28] is that the binder layer behaves like a photoconducting insulator. The active part in the binder system is the semiconductor powder. The role of the resin is merely to modify the surface states of these powders, thus influencing the nature and density of interface carrier traps in the binder system. Specific results emerging from the investigation include the following:

(i) The limiting discharge mechanism for dark decay in ZnO photoreceptors is the injection of surface charges into the bulk of the photoreceptor.

(ii) Similar dark decay experiments on CdS photoreceptors show that the discharge rate is limited by surface injection and bulk thermal generation of carriers.

(iii) The dependence of the dark discharge current on the resin concentration shows that for less than ~20 wt% of resin, the semiconductor grains are in contact with each other, enabling charge transport to occur between adjacent grains. For photoreceptors with more than ~30 wt% of resin, the dark discharge current decreases drastically because the semiconductor powders are separated from each other by a relatively thick layer of polymer binder influencing the charge transport between adjacent grains.

(iv) Analysis of photo-induced discharge characteristics for a step-function light exposure show that the electron ranges for ZnO and CdS photo-receptors are of the order of 10^{-5}–10^{-6} cm^2 V^{-1}.

(v) The residual voltage in these photoreceptors (containing ~15–20 wt% of resin) was measured to be unexpectedly large (−40 to −70 V for a typical sample 40 μm thick).

(vi) The transient photocurrent using short-duration photo-induced discharge techniques is always limited by the traps within the bulk of the photoreceptors.

(vii) Rose Bengal dye has a marked effect on the photoconductive properties of the ZnO photoreceptor. As expected, the dye extends the spectral response into the visible part of the spectrum.

(viii) The spectral photosensitivity of CdS powder layers can be enhanced to give a broad photoresponse in the visible region by heavy doping of copper and chlorine impurities.

4.3.3 Amorphous semiconductor photoreceptors

4.3.3.1 Charge carrier trapping in amorphous semiconductor photoreceptors

There are several experimental techniques for the investigation of charge carrier trapping in photoreceptors. The most common of these are:

(i) electroded time of flight (ETOF), in which the sample is sandwiched between two electrodes across which a voltage is applied [15];

(ii) xerographic time of flight (XTOF), where the top surface of the sample is floating and the applied electric field is achieved by electrostatic charging of the surface of the sample [29];

(iii) thermally stimulated discharge [30, 31] (see also pp. 4305, 4316 in the same issue as Reference 30);

(iv) *I–V* characteristics [32].

Among these experimental techniques, time of flight (ETOF or XTOF) is the most convenient and unambiguous method of determining the charge transport parameters μ and τ. Effects of impurities, applied electric field and temperature on μ and τ can be easily assessed by these techniques [33]. In this section, the principles of these measurements are illustrated by briefly examining the various forms of time-of-flight experiments. However, for more detail on these and other techniques mentioned above, the reader is referred to the appropriate references given at the end of this chapter.

The principles of ETOF and XTOF can be explained with the aid of Figures 4.9 and 4.10. In Figure 4.9, the device under test with thickness L is sandwiched between two electrodes. The top electrode is made transparent so that a short pulse of strongly absorbed light beam with duration t_{ex} excites a thin layer of carrier pairs. The wavelength of the light is chosen so that the absorption depth is very small compared to the sample thickness L. Depending on the polarity of the applied voltage V_0, either positive or negative carriers are drawn into the bulk. For the case shown, the electrons neutralise some of the positive charges on the top, whereas holes move towards the conductive substrate under the influence of the applied field $E_0 = V_0/L$. If the carrier lifetime is greater than their transit time across the sample (T_t), which is often the case, carriers will reach the opposite electrode and T_t will be given by

$$T_t = L^2/\mu V_0 \tag{8}$$

From the knowledge of T_t, V_0 and L, μ can be calculated.

The principles involved in XTOF are very similar except that the floating surface of the device resting on an earthed electrode is charged in

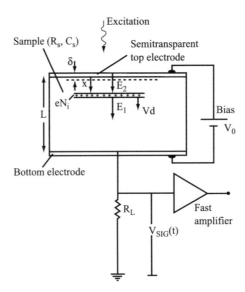

Figure 4.9   Illustration of the principles of ETOF

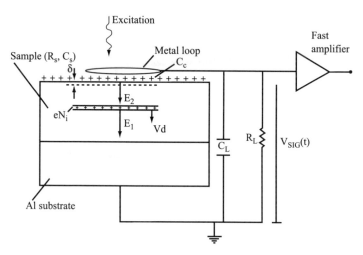

Figure 4.10   Illustration of the principles of XTOF

the dark to a positive surface potential (as shown in Figure 4.10) or a negative surface potential V_0 by a corona [23, 29]. In both experiments, if the photo-injected charge N_i is very small, the internal field is not distorted and $E_1 = E_2 = E_0$. This is known as a small-signal condition. As the generated carriers drift across the sample, the fields $E_1(x)$ and $E_2(x)$ cause a redistribution of the charge. The induced charge $q(t)$ on the bottom electrode at time t is

$$q(t) = N_i\, qx/L \qquad (9)$$

This gives rise to a current $I(t)$ to flow across R_L in the circuit of Figure 4.9. If the time constant of the circuit is smaller than T_t, it can easily be shown that [33]

$$V_{SIG}(t) \propto I(t) \qquad (10)$$

where

$$I(t) = q\, N_i\, /T_t;\; t \le T_t \qquad (11)$$

$$I(t) = 0;\; t \ge T_t \qquad (12)$$

This mode of operation is called the current mode of operation, which ideally would look like a square pulse. If the time constant of the circuit is larger than T_t, then

$$I(t) \propto \int I(t)\mathrm{d}t \qquad (13)$$

As the drifting charge sheet moves across the sample, if some of the carriers are removed from it due to trapping, then N_i in eqn. 11 exhibits a simple exponential density of the form

$$N_i(t) = N_0 \exp(-t/\tau) \qquad (14)$$

where τ is the deep trapping time, also known as the mean free time or carrier lifetime, and N_0 is the number of carriers created at $t = 0$. Thus with trapping the current $I(t)$ becomes

$$I(t) = [qN_0/T_t] \exp(-t/\tau) \qquad (15)$$

For illustration, Figures 4.11 and 4.12 show oscilloscope traces of typical XTOF hole photocurrent signals for a 58 µm pure a-Se sample [29] and a 58 µm 20 ppm Cl-doped a-Se + 5.3 wt% Te sample [11]. The signal in Figure 4.11 is a good example of single exponential decay satisfying eqn. 15. As expected, a change of the transparent top contact in ETOF does not seem to have a significant effect on τ. For example, Figure 4.13 shows the photocurrrent decay during the transit with Au, Al and Cu as different top contacts. The 10% variation in τ is not significant enough to draw any conclusion. Figure 4.12 is deliberately chosen to show another extreme case in which the signal exhibits a rapid initial decay in the form of a spike followed by a slow exponential decay. For calculation of the mean free time τ_h, the spike is neglected. A good explanation for the spike is significant fractionation in a-Se:Te alloys, resulting in a non-uniformly doped structure. This is discussed in more detail in the following section. Note that in Figures 4.11 and 4.12, the applied voltage V_0 is 115 V/58 µm in the a-Se sample and 200 V/58 µm in the Cl doped a-Se:Te sample. Using eqn. 8 with the known values of T_t, L and V_0, the hole drift mobility μ at the specified applied electric field can be calculated.

A practical ETOF/XTOF system developed by the author is shown in Figure 4.14. The major components of the system are:

   (i) an EHT supply, a corona device and a grid for charging the floating surface of the sample in the XTOF mode of operation;

  (ii) an electrostatic voltmeter with a transparent probe for measuring the potential applied to the sample in the XTOF mode of operation;

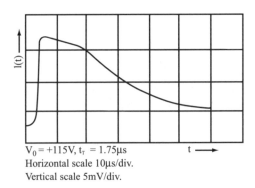

$V_0 = +115V$, $t_r = 1.75\mu s$
Horizontal scale 10µs/div.
Vertical scale 5mV/div.

Figure 4.11   Transient photocurrent signal in an a-Se photoreceptor [29]

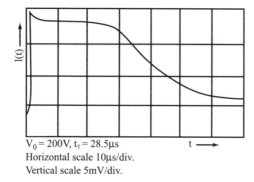

$V_0 = 200V$, $t_T = 28.5\mu s$
Horizontal scale 10μs/div.
Vertical scale 5mV/div.

Figure 4.12 Transient photocurrent waveform in a Cl doped a-Se:Te photoreceptor [11]

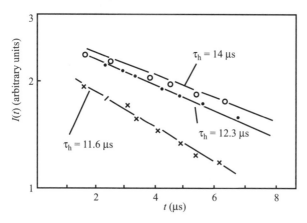

Figure 4.13 Determination of carrier lifetime in a-Se using ETOF and different top contacts of Au (×) with V = + 30 V, Al (•) with V = +20 V and Cu (○) with V = + 10 V

(iii) a power supply for applying voltage to the grid in the XTOF experiment or for applying a potential across the top and bottom electrodes in the ETOF experiment;

(iv) a xenon flash and a filter for photodischarge of the sample in the ETOF or XTOF experiments;

(v) a preamplifier (MA1) inside the sample box, and a main amplifier with a voltage gain of 10 for amplifying the ETOF or XTOF transient signal;

(vi) an oscilloscope for displaying the ETOF or XTOF signal;

(vii) a pulse generator, and a control unit for controlling events at various stages of the experiment.

In a typical XTOF experiment, the sample is charged at the charge position, it is then moved under the electrostatic position for detecting

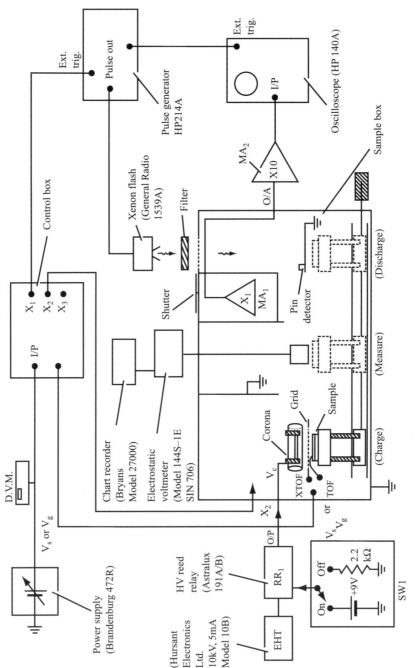

Figure 4.14   A schematic diagram for a practical ETOF/XTOF apparatus developed by the author

the applied voltage and finally it is brought to the discharge position for photodischarge by the xenon flash. In this mode of operation there are undesirable short delays between charging, detection and excitation depending on the speed of operation. However, these minor delays do not apply to the ETOF mode of operation because voltage is applied to the sample by a permanent metal wire contact to the transparent top electrode using conducting paste.

An advanced version of this system is shown in Figure 4.15 [34, 35]. The apparatus control and data acquisition system are operated by a combination of Turbo C and MATLAB programming [35]. The sample under test is positioned under the scorotron charging device which applies a user-defined surface charge of positive or negative potential to the sample. The scorotron is powered by two Brandenburg model 707R 15 kV dual polarity power supplies. An optical timing/position sensor (TPS1) unit is placed directly beneath the scorotron. A mechanical device is located beneath the sample, and is used to break a light beam generated by TPS1. Once the desired charge has been applied to the sample floating surface, the sample is moved automatically to a measuring station which consists of a second optical timing/position sensor unit (TPS2) and a Monroe 1009T transparent probe. The probe is used to

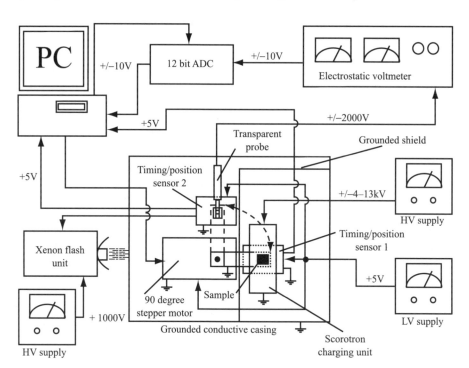

Figure 4.15 *A computer controlled experimental apparatus for charge transport measurements*

detect the applied charge and is a non-contact device which will drive to the potential under test in < 7 ms to an accuracy of 0.01%. For dark decay measurements, the decay curve is monitored via a Monroe model 144 electrostatic voltmeter and the developed data acquisition system. For photo-induced discharge measurements, the applied surface charge is discharged by a purpose built light source upon arrival at the measuring station. In both experiments, the data are automatically loaded into the MATLAB software package for analysis. The developed Turbo C program is responsible for the control of the experimental apparatus, which includes a stepper motor system, the timing position sensors system and the data acquisition system. MATLAB is used for data analysis and file management. The stepper motor and timing/position sensors system are controlled via the parallel port of the PC.

4.3.3.2   *Amorphous-selenium based photoreceptors*

Of the many amorphous semiconductor photoreceptors known to exist, amorphous selenium (a-Se) and its alloys are the most widely investigated [36–40]. These photoreceptors are commonly prepared by thermal evaporation on to an Al substrate, in the form of drums, and are machined, chemically cleaned and degenerated. a-Se and its alloys are then coated on them in a vacuum.

There are several factors to consider in this process: the coater design, particularly the layout of the mandrels that hold the drum, the substrate temperature and the crucible, which is usually a tube with a slot cut down through in a u or v shape. The mandrel layouts of the coater may be of any of the following three types: (a) parallel, in-line, with two to eight horizontal drum holders; (b) planetary, where the mandrels are arranged around the periphery of the chamber in a circular fashion, forming a cage; or (c) semi-planetary, where there are two small planetary units with the coater and a single crucible down under it. During the coating operation, the entire unit and each individual drum rotate.

For a-Se, a-Se:Te or a-Se:As, the substrate is heated to around 75°C whereas for a-As_2Se_3 this temperature is about 200°C. There are several methods for heating the substrate, for example using heated fluid running through the mandrel, glow discharge or ion bombardment or electron beam heating. Glow discharge or ion bombardment is used in many planetary systems, whereas electron beam heating is more common for use at higher deposition temperatures such as when evaporating a-As_2Se_3. Although there is no limit on the drum size, the larger the coated area the higher is the chance that a defect will occur when coating the drum. The function of the aluminium substrate is to support and position the photoconducting layer. For the photoreceptor to retain the surface charge in the dark, the opposite charge should not be injected from the substrate. As mentioned earlier, this is achieved by introducing a blocking layer between the substrate and the photoconductor to prevent injection. The most common solution to this is to grow an oxide layer on the substrate. Another method of preventing charge injection from the substrate is to use an additional intermediate layer with electronic properties that, when applied to the substrate, would build up a better barrier layer together with the photoreceptor. In the case of a-Se and its alloys, aluminium oxide is most widely used as the blocking layer.

From various experiments it is evident that the band-gap of a-Se is 2.1 ± 0.1 eV [41]. The models proposed for the electronic band structure of a-Se are typical of those suggested for amorphous semiconductors. While in crystalline semiconductors the effect of a suitable impurity is to provide a new donor or acceptor state, this is not always the case in amorphous semiconductors. Instead of providing a localised impurity level in the energy gap, an impurity may merely alter the mobility of charge carriers or it may introduce structural changes with or without modification of the localised states in the energy gap [1, 32, 42].

The quantum efficiency η in a-Se depends on the temperature, photon energy and applied electric field. It has been argued that η is controlled by the Onsager mechanism for the dissociation of an electron–hole pair generated by the same photon [43]. The initial electron–hole pairs generated by the absorbed photon remain mutually attracted by their coulombic interaction because they are not able to diffuse far enough apart during the thermalisation process. As the pairs diffuse in the amorphous structure, there is a certain probability that they will escape recombination. The applied field encourages the escape process, which gives η characteristics of S-shaped field dependence.

The optical band-gap of a-Se$_{1-x}$Te$_x$ alloys decreases as the Te concentration increases [18]. For example, for Te concentrations of 20 at%, 10 at% and 5 at%, the corresponding values of the energy gap are ~1.45 eV, ~1.55 eV and ~1.65 eV [1]. Both η and spectral photosensitivity have also been found to vary with the Te concentration [44]. The difficulty of producing homogeneous alloys of selenium has been known for some time. Severe fractionation of SeTe alloys has been observed with the surface tellurium concentration in the order of approximately double the base figure. Only the slope and plateau length alteration of Te concentration profiles are dependent on the temperature and initial Te concentration. Fractionation is only observed to the full predicted degree with an open source. In cases where the geometry is enclosed or near-enclosed, the vapour mixing evens out the Te concentration over the depth of the sample.

Various experiments suggest that in pure a-Se and its alloys, the electrical conductivity σ is thermally activated and satisfies the relation [32]

$$\sigma = C \exp(-E\delta/kT) \tag{16}$$

where C is a constant and $E\delta$ is the activation energy. The conductivity of pure a-Se is of the order of $10^{-16}\ \Omega^{-1}\ cm^{-1}$. The dependence of current or applied voltage resulting from injection of carriers into a-Se has been studied extensively. In most cases, the results obtained have shown that for small voltages conduction is ohmic but for high voltages, there is a non-linear relationship between the current density J and the applied voltage V. The current in this region has been interpreted as space-charge-limited due to the presence of trap levels in the energy gap. Depending on the distribution of the traps, the current-voltage relationship may take different forms [32, 33]. For example, for a uniform distribution of traps $J \sim AV \exp(BV)$, where A and B are constants, and for exponential distribution of traps $J \propto V^{3.8}$. For observation of dark space-charge-limited current (SCLC), at least one of the contacts to the sample must be injecting [33]. In general, ohmic contacts are used for this purpose. The current is said to be space-charge-limited when the density of injected free carriers exceeds the thermal equilibrium concentration of the free carriers in the sample. If the diffusion currents and effects of high electric field such as the Poole–Frenkel are neglected, the bulk space charge currents in a homogeneous medium satisfy the scaling law [32]

$$J/L = f(V/L^2) \tag{17}$$

where f is a function and L is the sample thickness. It should be noted that, in agreement with some other investigators, the author has found that experimental data on a-Se and its alloys cannot be correlated with the scaling law. The J–V characteristic curves do not always exhibit low electric field ohmic behaviour and $J \propto V^2$ regions are not always detected. Moreover, current densities have compatible magnitudes under both positive and negative applied voltages. Measurements at high electric fields seem to satisfy the relation

$$J = J_0 \exp(\gamma \sqrt{V}\ V) \tag{18}$$

Such a dependence may be attributed either to Schottky emission from the electrode or to the Poole–Frenkel effect. The above observations clearly indicate that more research should be conducted on the nature of metal–amorphous semiconductor contacts. As will be discussed in Section 4.4, understanding the physics of such contacts is essential for improving the performance of amorphous semiconductor X-ray image sensors.

The most common Se alloys used for manufacturing photoreceptors are pure Se, Se + 0.5 wt% As (referred to as stabilised Se), Se + 4–10 wt% Te and As_2Se_3. The major advantages and disadvantages of these photoreceptors are summarised in Table 4.4. In order to investigate the

effect of impurities, various a-Se based photoreceptors were manufactured by the author under carefully controlled fabrication conditions
such as the exact amount of added impurities, substrate temperature and
vacuum pressure.

Three experimental techniques, namely electroded time of flight
(ETOF), xerographic time of flight (XTOF) and xerographic discharge,
were used to determine the effects of impurity (As ~0.3 wt%, Te up to 17
wt%, Cl up to 60 ppm) on the drift mobility, carrier lifetime, dark decay
and residual voltage [11, 33, 44–46]. In general agreement with other
investigators, it was found that the addition of As or Te alone reduced
both hole and electron lifetimes. Both ETOF and XTOF experiments
showed a trap-limited response for electron transport in Se + Cl and
Se : Te + Cl photoreceptors. The addition of As was found to restore the
electron response in these devices. As shown in Figures 4.16 and 4.17, both
hole drift mobility μ_h (Figure 4.16) and electron drift mobility μ_e (Figure
4.17) in all Se-based systems are field dependent, satisfying the algebraic
relation, eqn. 19, where n is a constant varying from 0.03 (for electron
mobility in a-Se) to 0.47 (for electron mobility in a-Se + 17 wt% Te):

$$\mu \propto E^n \tag{19}$$

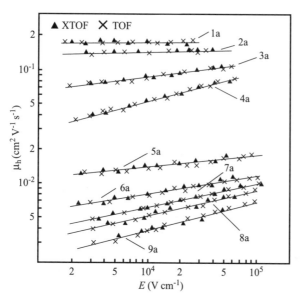

Figure 4.16 *Electric field dependence of XTOF and ETOF hole drift mobility in
various a-Se based photoreceptors: 1(a) 49 μm a-Se, 2(a) 48 μm a-
Se + 0.3% As, 3(a) 46 μm a-Se + 30 ppm Cl, 4(a) 66 μm a-Se + 17%
Te, 5(a) 39 μm a-Se + 0.3% As + 2.3% Te + 20 ppm Cl, 6(a) 68 μm
a-Se + 3.5% Te, 7(a) 60 μm a-Se + 5% Te, 8(a) 62 μm a-Se + 5.3%
Te + 16 ppm Cl, and 9(a) 62 μm a-Se + 12.5% Te + 20 ppm Cl*

Table 4.4   Advantages and disadvantages of photoreceptors based on a-Se and its alloys

Photoreceptor	Advantages	Disadvantages
a-Se	– Good sensitivity – Reusable – Good cyclic properties – Easy to manufacture	– Soft and scratches easily – Low surface crystallisation resistance – Sensitive to blue light only
a-Se + 0.5% As (stabilised Se)	– Harder surface than a-Se – More sensitive to crystallisation than a-Se	– Sensitive to blue light only
Arsenic triselenide	– Highest photoresponse – Fully panchromatic in visible light – Hard surface – Very high crystallisation resistance	– Difficult to manufacture – Expensive – Operating parameters sensitive to temperature changes – Exhibits cyclic fatigue
a-Se + 4–10% Te	– Higher sensitivity than a-Se and stabilised Se – Faster speed of response than a-Se and stabilised Se – Easy to manufacture – Inexpensive	– Less crystallisation resistance than a-Se and stabilised Se

Hole lifetime τ_h in a-Se:Te and Cl doped a-Se:Te photoreceptors is found to be field dependent in the form

$$\tau_h \propto E^k \tag{20}$$

where k is a constant. For the purposes of analysis and comparison of different photoreceptors, from the semi-logarithmic plot of τ_h vs E, the value of τ_h at zero field is obtained by extrapolation and taken as the hole lifetime. Evaluation of carrier lifetime from the transient photocurrent signals in conventional time-of-flight techniques (ETOF, XTOF) is limited to samples having low Te concentration (≤ 5 wt%). As discussed in Section 4.4, the advanced time-of-flight technique known as the interrupted transit time-of-flight technique (ITOF) can be employed in such cases [47].

Dark decay and residual voltage measurements were described in Section 4.2. Alloying a-Se with Te increases the dark decay, due to a bulk thermal generation process. Analysis of dark decay curves indicates that the volume density and energy spread of the mid-gap localised states in a-Se increases with the Te content. The repetition of the xerographic

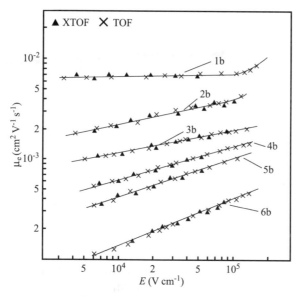

Figure 4.17 *Electric field dependence of XTOF and ETOF electron drift mobility in various a-Se based photoreceptors: 1(b) 59 μm a-Se, 2(b) 48 μm a-Se + 0.3% As, 3(b) 39 μm a-se + 0.3% As + 2.3% Te + 20 ppm Cl, 4(b) 68 μm a-Se + 3.5% Te, 5(b) 60 μm a-Se + 5% Te, 6(b) 66 μm a-Se + 17% Te*

cycles over a number of times leads to the saturation of the residual voltage which was used to calculate the concentration N_t of deep hole traps. The results show that the addition of Te increases N_t from ~1.9×10^{13} cm^{-3} for a-Se to 7×10^{13} cm^{-3} for a-Se + 8 wt% Te.

4.3.3.3 a-Si : H photoreceptors

One of the relatively new and popular types of amorphous semiconductor photoreceptor is hydrogenated amorphous silicon, a-Si : H. In order to achieve the desirable electrical properties, the hydrogen content of these photoreceptors is high, normally at 15–20 at%. The desirable characteristics are: good photosensitivity, long lifetime, high charge acceptance and slow dark decay [47–51]. a-Si photoreceptors are usually positively charged. To prevent the occurrence of electron injection from the substrate, a very thin B-doped layer, called a blocking layer, is normally placed adjacent to the substrate. The layer thickness and B concentration are selected such that electron displacement is much less than the layer thickness. Typically, the thickness and B concentration are 0.5–2 μm and 1000 ppm, respectively. Figure 4.18 shows schematically a typical photoreceptor structure, comprised of a few hundred

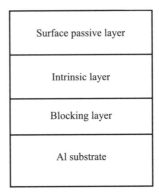

Figure 4.18   A typical structure for a-Si : H photoreceptor

Ångströms thick blocking layers adjacent to the aluminium substrate, an intrinsic layer of a few tens of microns thickness and a surface passivation layer a fraction of a micron thick. The intrinsic layer is normally very lightly doped with boron (10–50 ppm) and the surface passivation layer is either silicon carbide or silicon nitride. Room temperature electron and hole mobilities of a-Si are in the range of 0.5 and 10^{-3} cm²V⁻¹s⁻¹ [52]. The deep trapping lifetime for electrons is approximately 1 μs. For holes, values are typically 1 ms. The electron lifetime is considerably reduced with increasing B and similarly for holes via the inclusion of P.

The most widely used method of preparing a-Si : H is plasma-enhanced chemical vapour deposition (PECVD). Other techniques include thermal evaporation and sputtering. In PECVD, the Al substrate which is in the configuration of a drum is mounted within a vacuum chamber. In the most basic process, there are two electrodes: one is in the form of a cylinder, placed in proximity to the drum substrate, and the other is the substrate itself, which is normally biased negatively. At a vacuum of typically a few torr, a plasma is introduced between these electrodes by applying a potential of a few hundred volts. The feed gas, silane (SiH_4), is then fed into the chamber to interact with the plasma. Consequently, silane dissociates to positively charged ions and neutral radicals which are deposited on the drum surface. Other feed gases include disilane and silicon tetrafluoride. It is possible to mix silane with hydrogen, helium or argon. The film can be doped with either boron or phosphorus by simply mixing the feed gas with diborane or phosphine. It is therefore easy to change the composition of the feed gas to build a multi-layer photoreceptor. In this process as one deals with gas phase reactions which tend to be slow, the deposition rate is also very small. Depending on the silane concentration, the deposition rate can vary from a few microns per hour to a few tens of

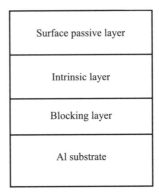

Surface passive layer

Intrinsic layer

Blocking layer

Al substrate

Figure 4.18   A typical structure for a-Si : H photoreceptor

microns per hour. The a-Si: H (p)/a-Si: H(I)/a-Si$_{1-x}$ C$_x$: H(p) structure is a good example of a commercial photoreceptor [52–54]. The total device thickness is typically 25–30 μm. The heavily doped a-Si: H(p) layer is ~0.5 μm thick. The a-Si : H(i)/a-Si$_{1-x}$C$_x$: H(p) layer is a heterojunction structure with a ~25 μm intrinsic layer of a-Si: H(i) with an energy gap of 1.72 eV and a fraction of a micrometre (0.2–0.3 μm) p-type a-Si$_{1-x}$ C$_x$: H(p) layer with an energy gap greater than 1.72 eV depending on the exact value of x.

Dark decay in a-Si: H photoreceptors is not well understood [40–43]. The dark decay of negative surface potential is injection limited and of no interest. The positive surface potential decays rapidly at the initial stage of the decay, followed by a slowly decaying residual voltage. It has been suggested that, within a certain time period, the dark decay is controlled by a Poole–Frenkel type emission of mobile holes and electrons [49]. Others propose completely different mechanisms [55–58]. One current research project involves investigation of the dark decay process in these photoreceptors.

Some of the major limitations associated with a-Si are cost, insensitivity to infra-red, high thermal generation and its significantly high capacitance, which leads to a substantial increase in the current required for the charging apparatus [3].

4.4 Amorphous semiconductor X-ray image sensors

Another important application of amorphous semiconductors is in digital X-ray imaging technology. The conventional method for obtaining X-ray images is based on analogue technology, which involves the use of a cassette containing film held in position behind an intensified phosphor screen. The screen absorbs X-rays, gives off light and exposes the film, which is subsequently processed to form a final image. Like other major medical imaging techniques such as magnetic resonance imaging, it is more desirable to employ digital radiography techniques. The benefits include greater precision of recording the information, increased flexibility of display characteristics and ease of transmitting images from one location to another over a communications network [59, 60].

Common digital radiography systems are based on phosphor screens or amorphous selenium [61, 62]. The use of a-Se for radiography has been known for many years but the efficiency and image quality had been limited in the past due to the toner development of the charge image. Several recent research groups have shown that combined with direct electrical readout a selenium detector can have advantages over conventional detectors with respect to signal-to-noise ratio and resolution [62–65].

A typical a-Se radiography system uses a 500 μm layer of a-Se containing ~0.5% As as an X-ray image receptor and is prepared by vacuum evaporation on to a metallic substrate. Application of an organic polymer layer of ~ 1 μm thickness on top of the selenium surface prevents reactions with the atmosphere. The whole process of X-ray image generation consists of three steps [62]:

(a) A homogeneous positive charge density is deposited on the Se surface using a corona charging device. Simultaneously, a negative bias voltage is applied to the aluminium substrate so that an electrical field builds up in the Se layer. A surface voltage of 1500 V results in an electric field of 3 kVmm^{-1} and an initial dark decay discharge of ~0.1 Vs^{-1}, which is adequate. Higher voltages can increase the sensitivity but they may cause electrostatic beakdown.

(b) The selenium layer is then exposed to X-rays. During exposure the surface charge is partially neutralised by the charge carriers which are created in the Se bulk and transported to the surface by the electrical field. The discharge is related to the local strength of the exposure. The residual surface charge distribution after exposure represents the X-ray image.

(c) The charge pattern is then read out non-destructively by microelectrometer probes which sense the surface potential through capacitive coupling. The probes scan the surface at a distance of ~ 100 μm. The electrical signal is amplified, digitised and fed into a computer for further processing, storage and display.

After readout the selenium layer can be recharged and used for the next exposure. This process erases the previous image; no separate erasing process is required.

a-Si:H has also been investigated for X-radiography [66–68]. When used for X-ray detection, a-Si:H cannot directly convert X-ray energy into electrical charge [69]. A scintillator is thus needed to do the conversion. One such sensor has been described as a 256×256 pixel array, with each pixel containing a photodiode and four switching diodes. The diodes were fabricated using plasma-enhanced chemical vapour deposition techniques [66]. The sensor employs $X-Y$ addressing of the rows and columns by the use of electrodes. The switching diodes are incorporated into the pixels to allow their individual addressing. An attempt has been made to devise a digital X-ray system based on a-SiN:H that does not require a scintillator to convert the X-ray radiation into visible light [68]. The a-SiN:H film is deposited on a 2 inch square glass substrate in a 100×100 array of thin film diodes of 200×200 μm dimensions. The system described requires no switching electronics and no X-ray conversion layer. The a-SiN:H is used to detect the emerging X-rays and also to generate an electrical signal.

There are many industrial process applications where digital radiography systems can be applied, such as crack, flaw and corrosion detection in processes where access to the media under test is impossible or difficult. For example, a system has been reported for imaging corrosion on pipelines which are covered in insulation and weather protection and are therefore inaccessible [70]. The system is a scanner-based design with a radiation source and a linear array of solid state photodiodes. A motor system moves the radiation source and sensor array along the section of pipeline to be imaged. Since in these applications the aim is to obtain quantitative data, the sensor array deployed is a simple 128 linear array of solid state detectors. The X-rays from the radiation source are passed through the pipeline under test. The emerging X-rays are converted to visible light by a scintillator and this light is detected by the solid state photodiodes array. Digital X-ray techniques also have applications in the food industry for the detection of foreign objects within foodstuff [71].

An advanced version of time-of-flight techniques known as interrupted transit time of flight (ITTOF) or interrupted field time of flight (IFTOF) has been found very useful for determining the nature of trapping kinetics in a-Se based photoreceptors [72] and a-Se : 0.5% As electroradiographic films used for X-ray imaging [73]. In ITTOF, the deep trapping characteristics are investigated with no electric field applied to the sample and at a point which is clearly in the bulk of the sample. In this respect, the time-of-flight methods described in Section 5.3.3 are inadequate for achieving this. For example, if conventional TOF is employed for investigating trapping in a-Se : 0.3% As X-ray medical imaging plates, the hole transit signal will be of a square shape showing no decay within the transit time, indicating that there is no deep trapping. As such, the signal provides no information on the trapping kinetics of holes to enable the evaluation of carrier lifetime or Schubweg for the sensor. On the other hand, in the ITTOF experiment, the interrupted photocurrent signal is considerably smaller than the pre-interrupted signal, thus providing information about the deep trapping [73]. Suppose that the field applied to the sample is removed for a period t_{int} after a delay t_d much less than transit time T_t, i.e. $t_d \ll T_t$. During the interruption, the carriers previously in transit are left to interact with the deep traps at that point in the sample which they had just reached prior to interruption of the field. If there exists an average characteristic lifetime for the carriers with respect to any deep traps, then the carrier density will decrease exponentially according to that lifetime τ. The number of carriers and hence the magnitude of the restarted current I_{res} will have decreased by a factor $\exp(-t_{int}/\tau)$ relative to the pre-interrupt value I_{pre},

$$I_{pre}/ I_{int} = \exp(-t_{int}/\tau) \qquad (21)$$

Both I_{pre}, the current just before interruption, and I_{int}, the current just after interruption, can easily be measured for different values of the interruption time t_{int}. A semi-logarithmic plot of $\ln(I_{pre}/I_{int})$ vs t_{int} gives a straight line, from the slope of which τ is found. Drift mobility is determined from the knowledge of transit time, applied field and sample thickness as in conventional TOF measurements.

The above analysis is just one of several applications of ITTOF. Other features of this technique include the following [72, 73]:

(i)   The time dependence of the instantaneous concentration of carriers in the transport band can be studied over a very long time scale, much greater than the transit time, until the recovered photocurrent signal is too small for detection.

(ii)   The effects of sample heterogeneities are eliminated by interrupting at a suitable time corresponding to a particular location in the sample.

(iii)   As a corollary to advantage (ii), sample inhomogeneities can be examined by interrupting the field while the photo-injected charge carrier packet is at different locations.

(iv)   The time dependence, if any, of the drift velocity, diffusion and dispersion can be also investigated.

4.5 Advances in the optical application of amorphous semiconductors

There have been considerable advances in the field of imaging technology [74–77]. Exploitation of this technology continues apace in facsimile, in desktop publishing, and in the other areas of business communication. As a result of continuous development, several optical techniques for writing on a photoreceptor have emerged [4, 78]. These methods may be divided into three groups of: (i) integrated arrays for parallel writing; (ii) cathode ray tube (CRT) for serial writing; and (iii) a laser deflector system for serial writing. For example, in (i) LED arrays have been used with a three layer Se–SeTe–Se photoreceptor. Magneto-optic light switching arrays have been fabricated on a single substrate, suitable for writing on many types of photoreceptors. The CRTs in (ii) offer reliability and total electronic control that is desirable for non-impact printing. Particular emphasis has been given to fibre optic CRT for high speed printing [78, 79]. Laser diodes in (iii) offer advantages of small size, low cost and self-modulating capability.

Once the optical method is chosen, appropriate photoreceptors more sensitive in the red and infra-red regions are required to write on. Organic materials, Te doped As_2Se_3, CdSe:Cu binder layers and a-Se based multi-layers have been reported as suitable photoconductors. Amorphous Se:Te alloys have been employed for laser colour printing. For example, high quality two-colour recording has been reported at a printing speed of 2730 lines per minute and a resolution of 240 dots per inch for a prototype printer equipped with an a-Se:Te photoreceptor sensitive in the long wavelength region and a diode laser with a wavelength of 780 nm and output power of 5 mW [80].

Multi-layer amorphous semiconductors have been found very useful for a number of applications such as TV image pickup tubes, erasable optical disks and ELICs (electrophotographic light-to-image converters) [78].

Figure 4.19 shows the simplified structure of a typical a-Se based TV pickup tube utilising the high panchromatic photosensitivity of the Se:Te alloy and the relatively fast hole drift mobility of a-Se. The electron beam injects electrons into the Sb_2S_3 layer and forms a negative space-charge region due to trapping. Layers of CeO_2 and Sb_2S_3 act as hole and electron blocking contacts, respectively. Photogenerated holes in a-Se travel across and recombine with the trapped electrons in the Sb_2S_3 layer. The photogeneration occurs in a region of high Te concentration.

A schematic diagram of an ELIC device is shown in Figure 4.20. In order to write and store the exposed image, a negative voltage is applied to the transparent ITO electrode. The image is converted to a stored negative space-charge region at the As_2Se_3 interface as follows. The ITO layer is exposed to the image and resulting photogenerated electrons transit towards the a-As_2Se_3 layer, where they are deeply trapped. To read the ELIC, the terminals are connected to a sampling resistor and the gold electrode side is scanned by a laser beam. Photogenerated holes are then driven by the electric field of the stored charge in order to neutralise the latter.

Because of the dielectric mismatch between various layers, drift mobility in a-Se based multi-layer structures is derived in terms of transit time, dielectric constant, thickness and applied voltage. Consider a typical double layer photoreceptor consiting of a top PGL and a second CTL.

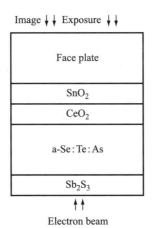

Image ↓↓ Exposure ↓↓

Face plate

SnO_2

CeO_2

a-Se:Te:As

Sb_2S_3

↑↑

Electron beam

Figure 4.19   A simplified structure for an a-Se based TV image pickup tube

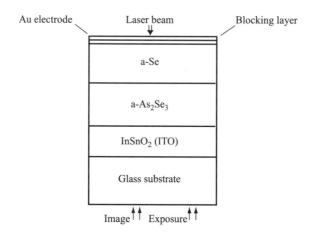

Figure 4.20 Example of a simplified structure for an ELIC device

Let the thickness, the relative dielectric constant, electric field and drift mobility be denoted by L, ε, E and μ with subscripts denoting the position of the layer (1 for PGL and 2 for the CTL). Assuming that the capacitance of each layer is constant, then from a simplified equivalent circuit for such a structure, it can be shown that the drift mobilities in the two layers are given by [81]

$$\mu_1 = [L_1^2 (1 - \varepsilon_1/\varepsilon_2) + L_1 L (\varepsilon_1/\varepsilon_2)]/[T_{r1} V_0] \tag{22}$$

$$\mu_2 = [L_2^2 (1 - \varepsilon_2/\varepsilon_1) + L_2 L (\varepsilon_2/\varepsilon_1)]/[T_{r2} V_0] \tag{23}$$

where T_{r1} and T_{r2} are transit times for the two layers. Note that if $\mu_1 < \mu_2$ the profile is of the form of a step due to the difference between the mobilities.

Practical transient current-signal profiles in multi-layer structures are not well understood. As an illustration, a typical low field signal indicating some of the defined regions is shown in Figure 4.21. The author has detected this kind of signal at low fields in various multi-layer structures. For both electrons and holes, at low applied voltage the features of the signal are: (i) an initial spike where the current rises to a peak value I_p and then decays exponentially to a smaller value; (ii) the spike which is followed by a relatively slow decay defined as the tail of the spike – at a transit time corresponding to the first layer, the tail of the spike is followed by a slowly rising portion called the saddle region of the signal; (iii) the growth of current terminates in a rounded peak I_2 at transit time T_{r2}. This peak is followed by a slowly decaying tail. As the applied voltage across the device is increased, the spike decay, the saddle rise time and the tail decay become very rapid, whereas the photocurrents I_1 and I_2 increase. For example, in Figure 4.22 the shape of the transient

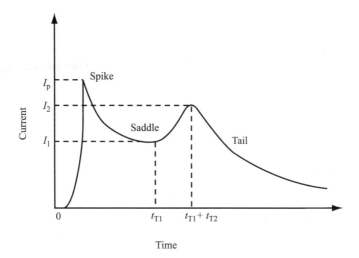

Figure 4.21   *Example of a low field transient photocurrent profile in a double layer structure with $\mu_1 < \mu_2$ showing various definitions*

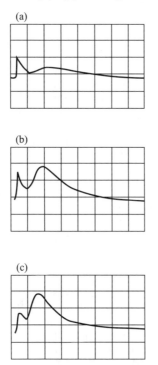

Figure 4.22   *ETOF transient photocurrent signals in a triple layer a-SeTe/Se/ SeTe photoreceptor showing the effect of applied field on the signal profile*

Horizontal scale 50μs/div, vertical scale is of an arbitrary unit

photocurrent signal in a triple layer a-SeTe(15%Te)/Se/SeTe(15%Te) changes significantly when the applied voltage is increased. The time scale for all three traces is the same(50 μs/div) but the applied voltage is –200 V in (a), –250 V in (b) and –300 V in (c). This interesting field dependent behaviour is not well understood but, as mentioned in Chapter 4.3.3, some of the observed features have been attributed to the fractionation and interface trapping [82, 83].

4.6 References

1   MOTT, N. F., and DAVIS, E. A.: 'Electronic processes in non-crystalline materials' (Clarendon Press, Oxford, UK, 1979, 2nd edn.)
2   ELLIOT, S. R.: 'Physics of amorphous solids' (Longman, London, UK, 1992)
3   BORSENBERGER, P. M., and WEISS, D. S.: 'Organic photoreceptors for xerography' (Marcel Dekker, New York, 1998)
4   SCHEIN L. B.: 'Electrophotography and development physics' (Laplacin Press, Morgan Hills, CA, USA, 1995)
5   MORT, J.: 'The anatomy of xerography' (McFarland, Jefferson, NC, 1989)
6   SCHAFFERT, R.: 'Electrophotography' (Focal, New York, 1980)
7   GAYNOR, J.: 'Advances in non-impact printing technologies for computer and office applications' 1982, The Society of Photographic Scientists and Engineers, Proceedings of the first international congress, Venice, Italy, June 1981 (Van Nostrand Reinhold Company, 1982)
8   SCHARFE, M.: 'Electrophotography principles and optimization' (Research Studies Press Limited, UK, 1984)
9   VAEZI-NEJAD, S. M.: 'Characteristics of pin and wire corona devices for electrostatics charging', *International Journal of Electronics*, 1989, **66**, (3), p. 437
10 JENNER, R. P., and VAEZI-NEJAD, S. M.: 'Development of instrumentation for a novel computer controlled experimental apparatus', Second international conference on *Planned maintenance, reliability and quality*, Oxford, UK, 2–3 April 1998
11 VAEZI-NEJAD, S. M., and JUHASZ, C.: 'Transient photoconductivity and residual voltage measurements on Cl doped a-Se : Te single layer and a-Se : Te/Se double layer semiconductor photoreceptors'. Institute of Physics Conference Series No. 89: Session 8 (IOP publishing Ltd, 1988), p. 283
12 VAEZI-NEJAD, S. M.: 'Xerographic discharge techniques for the investigation of charge transport in high resistivity materials', *International Journal of Electronics*, 1987, **62**, (2), p. 71
13 KASAP, S. O.: 'Charge-carrier deep-trapping kinetics in high resistivity semiconductors', *Journal of Physics D*, 1992, **25**, p. 83
14 BORSENBERGER, P. M., and WEISS, D. S.: 'Organic photoreceptors for imaging systems' (Marcel Dekker, New York, 1993)
15 VAEZI-NEJAD, S. M.: 'Novel experimental techniques for the evaluation of electrical properties of microelectronic materials'. Proceedings of the 1987 international symposium on *Microelectronics*, Minneapolis, USA, 28–30 Sept. 1987, p. 670

16 VAEZI-NEJAD, S. M., and JUHASZ, C.: 'Electrical properties of amorphous semiconductor selenium and its alloys, II – Multilayers', *Semiconductor Science and Technology*, 1988, **3**, p. 664
17 VAEZI-NEJAD, S. M.: 'Application of charge decay techniques tomicroelectronic materials and devices'. Proceedings of Nepcon Europe Conference, 20–22 March 1990, Birmingham, UK
18 JUHASZ, C., VAEZI-NEJAD, S. M., and KASAP, S. O.: 'Interface hole traps in double-layer amorphous semiconductors $Se_{1-x}Te_x$ photoreceptor devices', Letter to the Editor, *Semiconductor Science Technology*, 1986, **1**, p. 302
19 KASAP, S. O., BHATTACHARYYA, A., and LIANG, Z.: 'Decay of electrostatic surface potential on insulators via charge injection, transport and trapping', *Japanese Journal of Applied Physics*, 1992, **31**, p. 72
20 KALADE, Y. U., MONTRIMAS, E., and JANKAUSKAS., V.: Proceedings of the SIST, Springfield, VA, 1994, p. 747
21 GILL, D. W.: *Journal of Applied Physics*, 1979, **43**, p. 5052
22 BALTAZI, E. S.: *Journal of Applied Photographic Engineering*, 1980, **6**, p. 147
23 MELZ, P. J., *et al.*: 'Use of pyrazoline-based carrier transport layers in layered photoconductor systems for electro-photography', *Photographic Science and Engineering*, 1977, **21**, p. 73
24 BORSENBERGER, P. M. and WEISS, D. S.: 'Organic photorecepters for xerography' (Marcel Dekker, New York, 1988), chap. 10, p. 599
25 LAGOWSKI, J., *et al.*: 'Charge transfer in ZnO surfaces in the presence of photo-sensitizing rays', *Journal of Applied Physics*, 1978, **49**, p. 2821
26 BROICH, B., and HEILAND, G.: 'Charge transfer between ZnO crystals and dye layers', *Surface Science*, 1980, **92**, p. 247
27 LICHTENSTEIGER, M., and WEBB, C.: 'Corona discharge-induced surface chemical effects on II-VI compounds', *Applied Physics Letters*, 1981, **38**, p. 323
28 CHAN, Y. C.: 'Xerographic properties of CdS and ZnO semiconductor powder–resin binder layers', PhD thesis, Imperial College, University of London, 1983
29 VAEZI-NEJAD, S. M.: 'Xerographic time of flight experiment for the determination of drift mobility in high resistivity semiconductors', *International Journal of Electronics*, 1987, **62**, (2), p. 361
30 VAEZI-NEJAD, S. M., KAMARULZAMAN, B. M. Z., and JUHASZ, C.: 'Thermally stimulated discharge currents in a-Se:Te/Se double layer photoreceptors', *Journal of Materials Science* 1992, **16**, p. 4311
31 VAEZI-NEJAD, S. M., and JUHASZ, C.: 'Investigation of electrical properties of high resistivity materials for solid state sensors', Fifth IMCS, Special Issue of *Sensors and Actuators B*, 1994, **5**, p. 43
32 VAEZI-NEJAD, S. M.: 'Electrical properties of amorphous selenium alloys'. MSc dissertation, Brunel University, UK, 1981
33 VAEZI-NEJAD, S. M.: 'Electrical properties of selenium based amorphous multilayer xerographic photoreceptors', PhD thesis, Imperial College, University of London, 1984
34 VAEZI-NEJAD, S. M., and JENNER, R. P.: 'Principles and application of advanced transient photoconductivity techniques', IMTC Proceedings, vol. 2, 1997, p. 942

35 JENNER R. P., and VAEZI-NEJAD, S. M.: 'Investigation of a novel optical sensor for process tomography: I–Instrumentation'. Second international conference on *Planned maintenance, reliability and quality*, University of Oxford, April 1998

36 VAEZI-NEJAD, S. M., and JUHASZ, C.: 'Hole transport in selenium based amorphous xerographic photoreceptors', *Journal of Materials Science*, 1988, **23**, p. 3387

37 VAEZI-NEJAD,, S. M., and JUHASZ, C.: 'Effect of Cl on xerographic properties of a-Se:Te alloys, *Journal of Materials Science*, 1988, **23**, p. 3286

38 PAI, D. M.: 'Time-of-flight study of the compensation mechanism in Se-alloy', *Journal of Imaging Science and Technology*, 1997, **41**, p. 135

39 HOPPER, M. (Ed.): Proceedings of the twelfth international congress on *Digital printing technologies*. SIST, Springfield,VA, 1996, pp. 444, 451

40 PAI, D. M.: *Journal of Imaging Science and Technology*, 1997, **41**, p. 135; see also ANDERSON, J. (Ed.): Proceedings of eleventh international congress on *Advances in non-impact printing technologies*, SIST, Springfield, VA, 1995, p. 46

41 OWEN, A. E., and SPEAR, W. E.: 'Electronic properties and localised states in amorphous semiconductors', *Physics and Chemistry of Glasses*, 1976, **17**, (5), p. 174

42 ARSOVA, D., NEWHEVA, D., and VATEVA, E.: Proceedings of the eighth international school on *Condensed matter physics: electronic, optoelectronic, and magnetic thin films* (John Wiley, New York, 1995), p. 299

43 PAI, D. M., and SPRINGETT, S.: 'Physics of electrophotography', *Reviews of Modern Physics*, 1993, **65**, p. 163

44 CHEUNG, L. *et al.*: 'Selenium tellurium alloys as photoconductors', *Photographic Science and Engineering*, 1982, **26**, p. 245

45 VAEZI-NEJAD, S. M., and JUHASZ, C.: 'Electrical properties of amorphous semiconductor selenium and its alloys: I. Monolayers', *Semiconductor Science and Technology*, 1987, **12**, p. 809

46 JUHASZ, C., VAEZI-NEJAD, S. M., and KASAP, S. O.: 'Xerographic properties of single- and double-layer photoreceptors based on amorphous selenium–tellurium alloys', *Journal of Materials Science*, 1987, **22**, p. 2569

47 KASAP, S. O., POLISCHUK, B., and DODDS, D.: 'An interrupted field time-of-flight (IFTOF) technique in transient photoconductivity measurements', *Review of Scientific Instruments*, 1990, **26**, (8), p. 2080

48 VAEZI-NEJAD, S. M., BOTENNE, C., and MEHDIAN, M.: 'Computer modeling of electrical properties of amorphous semiconductor photoreceptors', International conference on *Concurrent engineering and electronic design automation*, Poole, UK, April 1994

49 STUZMAN, M.: *Applied Physics Letters*, 1991, **56**, p. 313

50 BAXENDALE, M., BISWAS, S., and JUHASZ, C.: SPIE, 1990, **1253**, p. 203

51 WAKITA, K., in YOKOYAMA, M. (Eds): Proceedings of the ninth international congress on *Advances in non-impact printing technologies*, SIST, Springfield, VA, 1993, p. 678

52 NAKAYAMA, Y., AKIYAMA, K., HAGA, N., and KAWAMURA, T.: 'Hole transport in glow-discharge produced a-Si H:F film', *Japanese Journal of Applied Physics*, 1984, **23**, p. L703

53 WAKITA, K., NAKAYAMA, Y., and KAWAMURA, T.: 'Preparation and

properties of GD a-Si films for electrophotography,' *Photographic Science and Engineering*, 1982, **26**, (4), p. 183

54 ASANO, A., ICHIMURA, T., OHSAWA, M., SAKAI, H., and UCHIDA, Y.: 'Characterization of a-Si/sub I-x/C/sub x/:H/a-Si: H and a-SiN/sub x/:H/ a-Si : H heterojunctions by photothermal deflection spectroscopy', *Journal of Non-Crystalline Solids*, 1987, **97–98**, p. 971

55 HIRAI, Y., KOMATSU, T., NAKAGAWA, K., and MISUMI, T., FUKUDA, T.: US Patent 4265991, 1981

56 NAKAYAMA, Y., KITA, H., TAKAHASHI, T., AKITA S., and KAWA-MURA, T.: 'Dark decay of surface potential: measurement of the density of localized states in highly resistive a-Si alloys', *Journal of Non-Crystalline Solids*, 1987, **97–98**, p. 743

57 PAASCHE, S. M., and BAUER, G. H.: *Materials Research Society Symposia Proceedings*, 1986, **70**, p. 671

58 PAASCHE, S. M., and BAUER, G. H.: 'Multilayered photo-receptors based on hydrogenated amorphous silicon', *Journal of Non-Crystalline Solids*, 1985, **77–78**, p. 1433

59 ZHAO, W., and ROWLANDS, J. A.: 'A large area solid state detector for radiology using amorphous selenium', Society of Photo-Optical Instrumentation Engineers, *Medical imaging VI: instrumentation*, 1992, 2676, p. 134

60 CHAN, H. P., DOI, K., GALHOTRA, S., VBORNY, C. J., MacMAHON, P. M., and JOKICH, P. M.: *Medical Physics*, 1987, **14**, p. 538

61 YAFFE, J., and ROWLANDS, J. A.: *Physics in Medicine and Biology*, 1997, **42**, p. 1

62 NEITZEL, U., MAACK, I., and GUENTHER-KOHLFAHL, S.: *Medical Physics*, 1994, **21**, p. 509

63 PAPIN, P. J., and HAUNG, H. K.: 'A prototype amorphous selenium imaging plate system for digital radiography', *Medical Physics*, 1987, **14**, p. 322

64 ROWLANDS, J. A., HUNTER, D. M., and ARAJ, N.: 'X-ray imaging using amorphous selenium: A photoinduced discharge readout method for digital radiography', *Medical Physics*, 1987, **18**, p. 421

65 ROWLANDS, J., and KASAP, S. O.: 'Amorphous semiconductor usher in digital X-ray imaging', *Physics Today*, 1997, (11), p. 24

66 GRAEVE, T., HUANG, W., and GUNTHER-KOHFAHL, S.: 'Image quality of a digital chest radiography system based on a selenium detector', *Society of Photo-Optical Instrumentation Engineers Journal*, **2415**, p. 177

67 APTE, R. B., STREET, R. A., READY, S. E., JARED, D. A., and MOORE, A. M.: 'Large area, low noise amorphous silicon imaging system', *Society of Photo-Optical Instrumentation Engineers Journal*, 1998, **3301**, p. 126

68 POPOV, I. A., VAN DOORSELAER, G., VAN CALSTER, A., DE-SMET, H., CALLENS, F., and BOEAMAN, E.: 'Prototype of 2D direct X-ray a-SiN:H sensor array', *Society of Photo-Optical Instrumentation Engineers Journal*, 1998, **3301**, p. 2

69 YAFFE, M. J., and ROWLANDS, J. A.: 'X-ray detectors for digital radiography', *Physics of Medical Biology*, 1997, **42**, p. 33

70 LOPUSHANSKY, R. L., LIGHT, G. M., and GUPTA, N. K.: 'Digital radiography of process piping', *Society of Photo-Optical Instrumentation Engineers Journal*, 1998, **3398**, p. 21

71 GRAVES, M., SMITH, A., and BATCHELOR, B.: 'Approaches to foreign body detection in foods', *Trends in Food Science & Technology*, 1998, **9**, p. 21

72 METHLEY, S. G.: 'Charge transport in amorphous Se:Te single and multilayer xerographic photoreceptor'. PhD thesis, University of London, UK, 1986, p. 86

73 POLISCHUK, B., and KASAP, S. O.: 'A high-voltage interrupted field time-of-flight transient photoconductivity apparatus', *Measurement Science & Technology*, 1991, **2**, p. 75; or see KASAP, S. O., POLISCHUK, B., and DOODS, D.: 'An interrupted field time-of-flight (IFTOF) transient photo-conductivity measurement', *Review of Scientific Instruments*, 1992, **61**, (8), p. 231

74 FLETCHER, M. J.: 'A review of optical inspection methods', *Insight – Non-Destructive Testing and Condition Monitoring*, 1996, **38**, (4), p. 36

75 WILLIAMS, R. A., and BECK, M. S.: 'Process tomography principles, techniques and applications' (Butterworth-Heinemann, Oxford, 1995), p. 5

76 HUANG, S.M., THORN, R., SNOWDEN, D., and BECK, M.S.: 'Tomographic imaging of industrial process equipment – design of capacitance sensing electronics for oil and gas based process', in 'Process tomography principles, techniques and applications' (Butterworth-Heinemann, Oxford, 1995)

77 HONLET, M.: 'Non destructive testing of complex materials and structures using optical techniques', *Insight – Non-Destructive Testing and Condition Monitoring*, 1998, **40**, (3), p. 176

78 DIAMOND, A. S. (Ed.): 'Handbook of imaging materials'. Diamond Research Corporation Ventura, CA. (Marcel Dekker, Inc., USA, 1991)

79 HAAS, W. H., GENOVESE, F. C., LANNOM, J. W., and WARNER, P. J.: 'High speed printing with fibreoptic-CRT Addressed Xerographic Engines', in 'Advances in non-impact printing technologies for computer and office application', The Society of Photographic Scientists and Engineers, *Proceedings of the first international congress*, Venice, Italy, June 1981 (Van Nostrand Reinhold Company, 1982) p. 412

80 HOSHI, N., KOMATSU, I., ANZAI, M., TANNO, K., MORISHITA, H., SAITO, S., and KATAGIRI, S.: 'Two-color electrophotographic printing method for laser beam printer', in 'Advances in non-impact printing technologies for computer and office application', The Society of Photographic Scientists and Engineers, *Proceedings of the first international congress*, Venice, Italy, June 1981 (Van Nostrand Reinhold Company, 1982) p. 518

81 VAEZI-NEJAD, S. M., and JUHASZ, C.: 'Electrical properties of amorphous semiconductor selenium and its alloys: II Multilayers', *Semiconductor Science and Technology*, 1988, **3**, p. 664

82 VAEZI-NEJAD, S. M.: 'Instrumentation aspects and application of charge transport measurement techniques', *Journal of the International Measurement Confederation (IMEKO)*, 1996, **17**, (4), p. 267–277

83 VAEZI-NEJAD, S. M.: 'Interpretation of transient photocurrent-time profiles in multilayer semi-insulators'. IEE fifth international conference on *Dielectric materials, measurements and applications*, University of Kent at Canterbury, UK, 27–30 June 1988

Chapter 5

Dielectrophoretic sensors for microbiological applications

D. W. E. Allsopp and W. B. Betts

5.1 Introduction

The microbiological world is both complex and diverse, impacting on almost every facet of human life in ways that are often poorly characterised, let alone understood. The scientific, social and economic impact of new and improved techniques for measuring the microbiological content of both natural and artificial environments could be immense. In this respect methods that enable the rapid, accurate and selective sensing of microbiological species will have enormous potential benefits.

Even within a relatively closed environment like a food processing plant a wide range of microbiological particles can exist, often in very small concentrations. Any method of monitoring or analysing the microbiological make-up of the sampled environment must overcome the problem of isolating a potentially low concentration of the particles of interest from an often diverse population. Such real world samples do not readily lend themselves to the biochemical or serological tests usually applied and the species under test must be pre-concentrated or selectively enriched, for example by bio-culture growth, prior to most chemically based analyses. Whilst these essentially chemical methods can be highly species specific, they are time consuming and inflexible, often requiring controlled environments or laboratory conditions.

Electronic sensors, or sensors that can be readily interfaced with electronic instruments, in principle offer the advantages of speed, high sensitivity (for example, impedance changes of 1 part in 10^9 can be detected), robustness and flexibility. Yet any practical electronically based microbiological sensor must be able to pre-concentrate, detect, respond selectively to and hopefully quantify the particles of interest. Devices based on the electromagnetic phenomenon known as *dielectrophoresis* can fulfil these requirements and in principle provide a basis for sensing in a wide range of microbiological applications [1, 2]. Dielectrophoresis results from the interaction between a polarised but otherwise neutral particle and an applied electric field that has induced the polarisation. In this chapter the basic physical principles of dielectrophoresis and its application to sensing microscopic particles are described. As will be seen, the technique requires the application of high ac electric fields between two or more electrodes. An overview of the design and fabrication, using conventional semiconductor device fabrication techniques of electrode cells suitable for microbiological sensing, is presented, along with details of measurement systems. Finally, several applications of dielectrophoretic sensing to microbiology are reviewed.

5.2 Fundamentals of dielectrophoresis

Dielectrophoresis is the term used to describe the polarisation and associated motion induced in otherwise uncharged particles by an ac electric field. The phenomenon arises from the difference in the magnitude of the force experienced by the charges at each end of the induced dipole when a non-uniform electric field is applied. Following the analysis of Pohl [1], it is useful to compare the different responses of charged particles and electrically neutral particles to uniform and non-uniform electric fields.

Figure 5.1a shows the case of charged and neutral particles in a uniform electric field. The ionised particle (if positively charged) drifts in the direction of the applied field. On the other hand the neutral particle is merely polarised. Whilst under certain circumstances a rotational motion may be induced, tending to align the dipole with the field, the neutral particle does not experience a net translational force that impels it towards either electrode.

When a non-uniform electric field, of the type shown in Figure 5.1b, is applied, the behaviour of a neutral, polarisable particle differs from that of the charged particle, which still drifts along the field lines to the electrode of appropriate polarity. In the case of the polarised but otherwise neutral particle the detailed interaction between the applied field and the charges forming the dipole must be considered. For the situation shown in Figure 5.1b, the negative pole induced in the external field experiences a stronger force (indicated by the closer spacing of the field lines) than the positive pole, despite the net charges in both poles being equal and opposite. As a result there is a net force on the polarised, neutral particle, usually moving it towards regions of greater field strength. This force is known as the dielectrophoretic force, and the translational motion of the neutral particle via the combination of

(a)

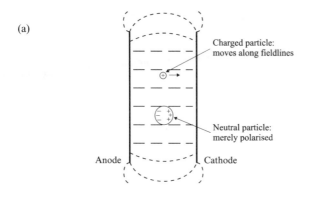

(b)

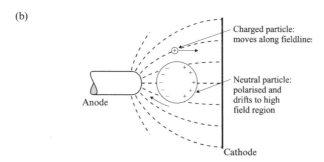

Figure 5.1 *The difference between the behaviour of neutral and charged particles in (a) a uniform electric field and (b) a non-uniform electric field (see Reference 1)*

induced polarisation and non-uniform electric field is known as dielectrophoresis.

The dielectrophoretic force depends critically on the net polarisability of the particle relative to the surrounding medium. As such it provides a means of detecting the presence of polarisable particles, of collecting these particles in regions of either low or high electric field gradient and, in many cases, of 'fingerprinting' the particles via their characteristic response to dielectrophoretic forces. These properties of dielectrophoresis become more apparent from the following analysis, which is based on that of Pohl [1].

The magnitude of the dielectrophoretic force can be calculated by considering separately the forces arising from a non-uniform electric field acting on the positive and negative poles of the induced dipole and which is orientated along the x-axis, as shown in Figure 5.2. If the electric field vector is given by $\mathbf{E} = (E_x, 0, 0)$, the magnitude of the force on the negative pole located at co-ordinate x_1 and having charge $-q$ is given by

$$F^- = -qE_x(x_1) \qquad (1)$$

Similarly, the magnitude of the force acting on the positive pole of charge $+q$ at position x_2 is

$$F^+ = qE_x(x_2) \tag{2}$$

Since the particle is small the electric field variation over its length, h, can be expressed as a Taylor series, where to first order

$$E_x(x_2) = E_x(x_1) + h\frac{dE_x}{dx} \tag{3}$$

Now the strength of the dielectrophoretic force is just the sum of F^+ and F^-, making

$$F_{DE} = F^+ + F^- = qh\frac{dE_x}{dx}$$

Equation 4 reveals that F_{DE} is proportional to the gradient in the electric field. As a consequence, for electrode geometries of the type shown in Figure 5.1b the particles undergo translational motion towards the region of highest electric field gradient that occurs at the pointed electrode. It is now shown that the aggregation of particles at the point of highest field gradient occurs, irrespective of the direction of the electric field, by considering the dipole moment induced in a particle subjected to a dielectrophoretic force.

The induced dipole moment is a vector quantity, the amplitude of which is related to the external electric field strength via

$$|\mathbf{p}| = p_x = qh = \chi v E_x \tag{5}$$

where χ is the susceptibility to polarisation per unit volume of the particle, v is the particle volume and E_x is the applied electric field strength. Combining eqns. 4 and 5 yields the result

$$F_{DE} = \chi v E_x \frac{dE_x}{dx} = \frac{1}{2}\left(\chi v \frac{dE_x^2}{dx}\right) \tag{6}$$

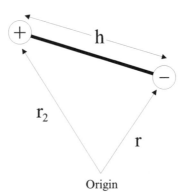

Figure 5.2   One dimensional dipole formed in a non-uniform electric field

Equation 6 shows that F_{DE} depends on E_x^2, making it independent of the direction of the electric field, as already mentioned. Put another way, the sign of the polarisation and of the field gradient change identically in eqn. 6, making the aggregation of polarised, neutral particles in regions of high electric field unaffected by the periodic reversal of the ac field.

Equation 6 is also recognisable as the x-component of a dielectrophoretic force deriving from a generally directed electric field given by $\mathbf{E} = (E_x, E_y, E_z)$ that gives rise to an induced dipole moment $\mathbf{p} = (p_x, p_y, p_z)$ on the neutral particle. Extending the analysis to a three dimensional coordinate space yields

$$\mathbf{F}_{DE} = \left(p_x \frac{\partial}{\partial x}, p_y \frac{\partial}{\partial y}, p_z \frac{\partial}{\partial z} \right) \mathbf{E} = (\mathbf{p}.\nabla)\mathbf{E} \tag{7}$$

Noting from eqn. 5 that $\mathbf{p}$ depends on the polarisability of the particle and the electric field, the dielectrophoretic force can be rewritten in the following form:

$$\mathbf{F}_{DE} = (\mathbf{p}.\varDelta)\mathbf{E} = \chi v (\mathbf{E}.\ \nabla)\mathbf{E} \tag{8}$$

Equation 8 reveals that if χ, the net susceptibility to polarisation per unit volume of the particle, is a positive quantity, then the dielectrophoretic force increases as both the electric field strength and the gradient in the field increase. As a consequence, the aggregration of particles occurs at regions where the electric field strength and the gradient in the field are simultaneously a maximum when they are subjected to such a positive dielectrophoretic force.

It should be noted that χ is strictly a tensor quantity with the magnitude and sign of its elements depending on the permittivities and conductivities of the particle and the suspending medium [3]. In the above analysis it has been assumed that the polarisability of the particles under consideration exceeds that of the medium in which they are suspended; but such circumstances do not always prevail. If instead the polarisability of the suspending medium is greater than that of the particles, then the dielectrophoretic force on the latter is strongest in regions of *lowest* electric field gradient, with the result that particles are induced to move away from electrode edges where the electric field is usually high. This phenomenon is known as *negative* dielectrophoresis [3] and, with a suitable choice of electrode geometry, can be exploited to entrap microscopic particles in spaces well away from the electrode edges [4]. Finally, it should be noted that susceptibility to polarisation is in general a complex quantity. As a consequence, significant phase differences between the diectrophoretic force and a driving ac electric field can be significant and can contribute to the induced particle motion [5, 6].

5.3 Sensing by dielectrophoresis

Having established the physical basis of dielectrophoresis, its application to sensing can be demonstrated by considering the electrode structure shown schematically in Figure 5.3. In the case of co-planar electrodes the gradient in the electric field is highest at the inner edges of the conducting strips (Figure 5.3a), irrespective of whether ac or dc voltages are applied. When a suspension of neutral particles is introduced to the space above the electrodes and an ac voltage is applied, particles are first polarised then, if the permittivity of the medium is lower than that of the particles, the latter rapidly gather at the regions of high electric field gradient under the influence of a positive dielectrophoretic force (Figure 5.3b). If the particles are large enough, the collection by the electrodes can be observed by an optical microscope.

Figure 5.4 shows an example of collection by dielectrophoresis. Here yeast cells are extracted from an aqueous suspension when an ac voltage is applied to co-planar aluminium electrodes of the type shown in Figure 5.3. As expected, the yeast cells initially collect at the inner edges of the electrodes where the field gradient is largest. As the number of particles aggregated in the high field regions increases, the dipoles induced in neighbouring particles interact to form what are known as pearl-chains

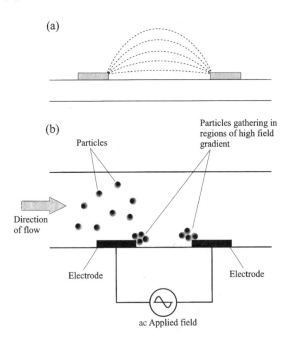

Figure 5.3   *(a) Schematic electric field lines emanating from co-planar electrodes and (b) dielectrophoretic collection of particles in suspension in a medium of lower permittivity*

bridging the electrode gaps [1, 2]. Pearl-chains appear in Figure 5.4 as lines of yeast particles stretching out into the electrode gap, but in reality the chains lie along the field lines arching from electrode to electrode.

Whilst the collection of particles is useful in itself in assisting micro-biological testing, for example in pre-concentrating a species in dilute suspension prior to further analysis, dielectrophoresis has further attrib-utes that are potentially advantageous for sensing. From eqns. 6 and 7 it can be seen that the strength of the dielectrophoretic force depends on the polarisability of the particles, which in turn depends on the frequency of the applied ac field. The dipole induced in a particle will not necessar-ily reverse itself in phase with an ac drive field, since charges in the poles may not redistribute rapidly enough as the field changes sign at higher frequencies, owing to the limited mobility of the charges making up the dipole and the frequency dependence of the real part of the permittivity of the particle. As a result, the strength of the instantaneously induced dipole moment and its time dependence will be functions of frequency, as well as the magnitude and direction of the applied field.

From the above discussion it can be seen that a device based on dielectro-phoresis in principle will have the desired characteristics for rapid, on-line microbiological sensing. However, exploiting dielectrophoresis in a prac-tical sensor is not straightforward and attention must be paid to monitor-ing the collection process, its quantification and to the control of factors such as the conductivity of the medium containing the particles. In essence, a whole sensor system involving at least some electronic instru-mentation is required. Some of these systems are discussed in the next section.

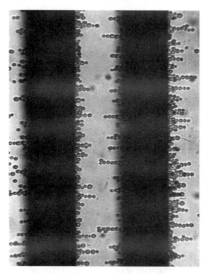

Figure 5.4   Dielectrophoretic collection of yeast cells from aqueous suspension at aluminium electrodes

5.4 Dielectrophoretic sensor systems

The heart of any dielectrophoretic sensor system will be formed by the electrodes from which the ac driving field is applied. The dependence of dielectrophoretic force on both the gradient and strength of the applied electric field means that the electrode configuration needed for efficient dielectrophoretic collection should produce strong, highly non-uniform fields if positive dielectrophoresis is to be used. On the other hand, if negative dielectrophoresis is to be exploited the electrode geometry should produce regions where the electric field gradient and E^2 are low [4]. In both cases the required electrode configurations can be realised straightforwardly using planar microfabrication processes adapted from silicon device technology with a suitable choice of substrate.

Figure 5.3 shows schematically what is perhaps the simplest electrode structure that has been used successfully in microbiological applications of dielectrophoresis [7]. From straightforward electrostatic consider-ations it can be seen that the electric field has a maximum in both its gradient and its strength at the inner upper corners of the electrodes. Such structures can be made using the standard microfabrication tech-nique known as lift-off to define electrodes having the necessary sharp edges. The method is shown schematically in Figure 5.5.

A substrate (typically glass) is first coated with a photoresist layer in which the electrode pattern is defined following exposure and develop-ment (Figure 5.5a). The electrode pattern is in effect formed by a sequence of windows in the photoresist through which a metal film is deposited on to the substrate either by evaporation or by sputtering (Figure 5.5b). The photoresist is then removed by a suitable solvent,

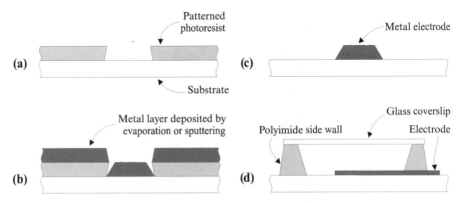

Figure 5.5   Lift-off method of forming co-planar electrodes: (a) substrate coated with patterned photoresist; (b) deposition of metal layer; (c) removal of photoresist and areas of unwanted metal; and (d) final electrode cell after formation of polyimide walls and fixing of a glass coverslip

simultaneously lifting off the unwanted areas of metal and leaving intact the desired electrode pattern (Figure 5.5c).

As most microbiological particles of interest can be analysed conveniently in suspension in a liquid medium, the substrate must either then be placed within a suitable container or, better still, some form of containment must be integrated on to the substrate. The latter can be achieved by using polyimide, which can be readily patterned by photolithography, to form impermeable walls surrounding all but the contact pads of the electrodes. A glass coverslip or slide can then be glued over the whole structure to form an enclosed dielectrophoretic sensor (Figure 5.5d). The suspension of particles is introduced into the electrode cell either via a syringe inserted through the polyimide walls, with a corresponding outlet through the side wall, or by inlet and outlet tubes fixed in the coverslip, as shown in Figure 5.6.

The detailed arrangement of the electrodes and their shape will be determined by the application and the type of dielectrophoresis to be exploited and the method of detecting particle motion. Optimal electrode geometries for particle collection can be found by numerical simulation to identify regions of maximum potential energy (for positive dielectrophoresis) and minimum potential energy (for negative dielectrophoresis) of the particles [3, 4, 8]. The procedure, detailed in Reference 4, involves calculating the electric field distribution for a given electrode from numerical solution of Poisson's equation, or its integral equivalent, and then determining the time averaged potential energy of a prototype particle.

For negative dielectrophoresis the electrode configuration ideally should form zones where the electric field gradient is low and in which particles become trapped [4]. Figure 5.7 shows schematically the formation of a low field trap by using four electrodes with adjacent electrodes

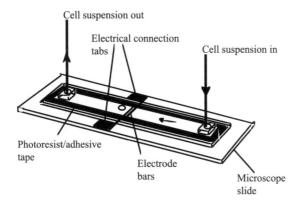

Figure 5.6   Complete dielectrophoretic electrode chamber. The internal height of the chamber is about 150 μm

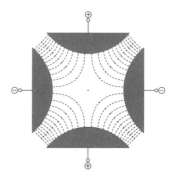

*Figure 5.7   Potential well created by four electrodes with adjacent electrodes
biased with positive or negative voltage*

biased with voltages of opposite polarity [5, 9]. The dielectric properties
of the medium and particles are adjusted, or the signal frequency chosen,
so that the particles are subject to a negative dielectrophoretic force. The
phase of the voltages applied to the electrodes can be varied to create a
rotating electric field which, in turn, causes the trapped particles to levi-
tate or to rotate at a characteristic rate [9, 10]. More complex configurations of up to eight electrodes have been utilised to exert greater control
of the rotating field in such a potential trap [9].

Since particles can be levitated as well as impelled towards regions of
low or high electric field gradient, the vertical as well as horizontal electric field distribution and its time dependence must be considered too.
Figure 5.8 shows the calculated time averaged potential energy profile for
a 3 μm radius particle suspended in an aqueous medium at a height
of 3.5 μm above the surface of the electrodes experiencing a positive
dielectrophoretic force that results from applying an ac voltage to the
interdigitated electrodes shown in the left-hand part of the figure [4].

The gap size between co-planar electrodes and hence the electric field
strength attainable from a given ac voltage generator is limited to about
1 μm if conventional photolithography is used. Müller and co-workers
[9] report the fabrication of co-planar quadrupole electrode structures
with gaps down to ~0.5 μm formed by electron beam lithography. By
using electrodes with such geometries, these authors were able to detect
dielectrophoretic collection of fluorescent latex particles with diameters
down to 14 nm, a size much smaller than the diffraction limit of resolution of optical instruments. Collection was monitored by observing
with a fluorescence detecting microscope the light re-emitted by the
particles when excited by shorter wavelength illumination.

For particles of diameter ≥ ~1 μm dielectrophoretic collection can be
detected optically using a microscope with the type of planar electrode
cell shown in Figure 5.6. A CCD camera and monitor linked to a frame
store enable image capture for particle counting by image analysis, to

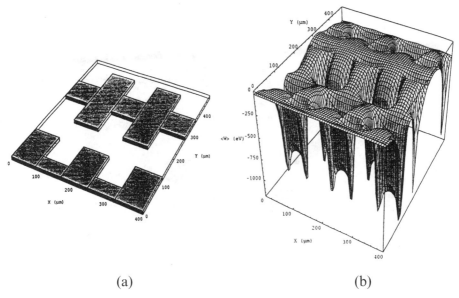

(a)   (b)

*Figure 5.8   Time averaged potential energy distribution (b) for a 3 μm radius
particle suspended in water at a height of 3.5 μm above the surface of
interdigitated, castellated electrodes (a) whilst subject to a positive
dielectrophoretic force. Data reproduced by permission from Wang
et al. [4]*

quantify the variation of collection efficiency of the particle under
test with the electric field strength and frequency. Figure 5.9 shows schematically a computer controlled system that enables dielectrophoretic
collection efficiency to be measured for user-defined applied ac voltages
and frequency range [11].

In this system dielectrophoretic collection is quantified by digitising
the image of an area downstream from the electrodes and counting the
particles released after the electric field is removed. After particle
release and counting, the peristaltic pump is triggered to flush out the
electrode cell and pump in a fresh volume of the suspension, and the
signal generator is triggered to repeat the measurement cycle either
under the same conditions or with a change in the signal frequency or
amplitude. The pump speed of the peristaltic pump is varied according
to the stage in the measurement cycle, to enable sequential flushing of
the electrode cell and refreshing of the suspension from which the particles are collected. By this method averaged plots of the dielectrophoretic collection efficiency of the type described in the next section
can be obtained.

Another feature of the dielectrophoretic sensor system in Figure 5.9 is

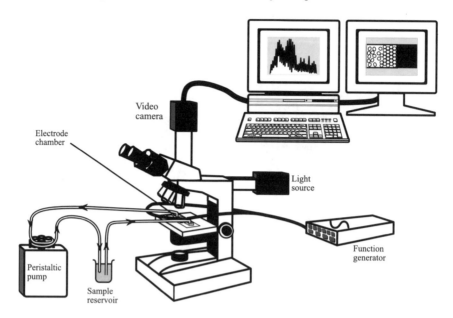

Figure 5.9   Dielectrophoretic sensing system using a CCD camera to detect particle collection. The system is computer controlled

the inclusion of pH meters before and after the electrode cell (not shown, for clarity), to measure the conductivity of the suspension. The pH meters are required for quantitative measurements of the frequency dependence of dielectrophoretic collection because the conductivity of the medium affects the observed response of a given particle to a dielectrophoretic force [1]. The pH meter downstream of the electrode chamber enables monitoring of any changes of the conductivity arising from damage to the microbiological cells caused by the polarising electric field. Some cells can be easily damaged by the ac electric field, releasing ions into the medium, or their collection characteristics can be susceptible even to small changes in the medium conductivity. If the conductivity does change then the field strength or its frequency can be adjusted to minimise the risk of cell damage.

The combination of a high power microscope, a CCD camera and a frame store provides a versatile basis for investigating dielectrophoretic forces acting on small particles. A system has even been developed for analysing the behaviour of a single microbiological cell subject to a positive dielectrophoretic force [12]. In this method the suspension is introduced into a chamber containing a downward pointing, conic electrode above a horizontal, planar electrode. The electrode configuration is essentially that shown in Figure 5.1b but rotated through 90°. The electric field lines are vertically aligned and the dielectrophoretic force impels any particle held in the electrode gap upwards towards the region of high field gradient, at the point of the upper electrode, against the gravitational force. A light source and CCD camera mounted on the horizontal axis and aligned to the electrode gap detects the vertical and lateral motion of the particle. The voltage applied to the electrodes is then adjusted via fast feedback control so that the oppositely directed dielectrophoretic and gravitational forces just balance. The dielectric properties of the trapped particle can then be analysed by varying the frequency of the applied voltage and monitoring the amplitude change needed to maintain it in a steady position [12].

The above methods are unsuitable for investigating the dielectrophoretic response of particles smaller than the resolution limit of the optical system imposed by diffraction. In practical terms this prevents their use in studies of particles significantly less than ~0.5 μm in extent. As many microbiological particles of current interest can be much smaller, particularly viruses and DNA fragments, other methods of sensing dielectrophoresis are required. Dielectrophoretic collection of fluorescently labelled particles and viruses of length less than 100 nm has been observed using a fluorescence detecting microscope [9, 13, 14]. Although this technique is not quantitative it is effective in assessing whether particles are subject to a positive or negative dielectrophoretic force and the onset of collection [14].

An alternative and potentially more powerful approach to detecting dielectrophoresis of nanoscale particles is to measure the change in low frequency electrical impedance between the electrodes as collection occurs. Since the dielectrophoretic force arises from the difference in the complex permittivity between the particles and the suspending medium, it is evident that there must be a change in the impedance between the electrodes as particles collect. With reference to Figure 5.3b it can be seen that if n particles of volume v collect from a dilute suspension on electrodes, contained in an effective volume V, the change in complex permittivity about the electrodes will scale with the volume of the medium replaced by particles according to[1]

$$\Delta \varepsilon^* \approx \frac{nv}{V} (\varepsilon_p^* - \varepsilon_m^*) \tag{9}$$

where ε_p^* is the complex permittivity of the particle and ε_m^* is the complex permittivity of the medium.

At low frequencies the impedance between the electrodes behaves as a capacitor shunted by a resistor and the change in complex permittivity affects mainly the capacitance through the variation in dielectric

[1] ALLSOPP, D. W. E., MILNER, K. R., BROWN, A. P., and BETTS, W. B.: 'Measurement of dielectrophoretic collection by impedance analysis', submitted to *Journal of Physics D*, 1998

constant. Thus, from eqn. 9, determining the capacitance change at low frequency provides a potentially quantitative measurement of the number of particles collected. Figure 5.10 shows an ac bridge circuit developed for measuring low frequency impedance changes arising from dielectrophoretic collection [15].

The system makes use of two electrode chambers integrated on to the same glass substrate and fabricated by the techniques described earlier in this section. The integrated chambers are then connected into the separate arms of an ac Wheatstone bridge. A small amplitude, low frequency probe voltage, V_p, is inductively coupled to the input of the bridge and applied simultaneously with the higher frequency voltage which drives the dielectrophoretic force. A particle suspension is pumped through one chamber, whilst the suspending medium alone is pumped through the second, to enable any impedance change arising from the polarisation of the medium to be accounted for. Capacitors C_1 in each arm of the bridge act as a high frequency short circuit for the resistors R_1, so that the large drive voltage is dropped across only the electrode chambers. The frequency, ω_p, of the probe signal is chosen to make $\omega_p C_1 \ll R_1$ so that the impedance change arising from dielectrophoretic collection can be determined from the difference voltage at this frequency measured by the lock-in amplifier. The frequency and amplitude of the drive voltage can be varied whilst V_p and ω_p are held constant.

Figure 5.11 shows a comparison of the dielectrophoretic collection spectra of the bacterium *B. subtillis* measured by the image analysis shown in Figure 5.9 [11] and by the impedance change method for the same suspended particle concentrations [15]. Another advantage

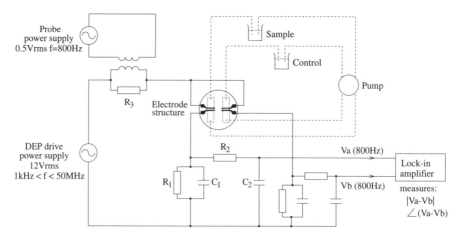

Figure 5.10   AC bridge circuit for measuring the low frequency impedance change arising from dielectrophoretic collection of particles from suspension

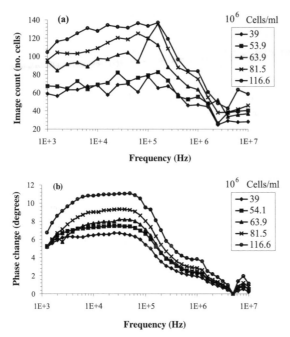

Figure 5.11   *The frequency dependence of (a) the number of* B. subtillis *counted measured by image analysis technique and (b) the corresponding change in the phase of the probe voltage dropped across the electrodes*

of measuring dielectrophoretic collection by impedance change becomes apparent. The point-to-point variations are much smaller for the impedance change method and, as a result, changes in collection efficiency with frequency that are possibly characteristic of the particle type become more clearly resolved. In addition, compared with the image analysis method, the time taken to complete a spectrum is halved, thereby enabling more rapid assessment of samples – an important consideration in possible clinical or biomedical applications of dielectrophoresis.

Unlike optical systems, there is in principle no limit on the particle size that can be detected using the impedance change method. For example, measurements of the dielectrophoretic collection spectra of 20 nm diameter latex beads have been reported, in which the amplitude of a region of characteristic impedance change scales with increasing particle concentration in suspension [16]. Also the particles under test do not have to be fluorescently labelled, thus eliminating a procedure that can alter the biochemical or biophysical properties of microbiological cells.

The coplanar electrode chambers discussed so far have the advantages of easy integration with other instrumentation to form a complete sensor and of cheap, straightforward fabrication technology. Their disadvantage lies in that only limited specimen volumes can be sampled, as typical cell volumes are only ~50–500 µl. To overcome this limitation, grid electrodes that allow sample flow perpendicular to the plane of the electrodes over a reasonable cross-sectional area have been developed and used successfully to detect and collect microbiological cells from large volumes of very dilute suspensions [17, 18]. The schematic layout of a grid electrode cell is shown in Figure 5.12. Unlike the planar structures the electrode geometry now prevents straightforward, direct visual sensing of dielectrophoretic collection. Instead it is measured by a change in optical absorption of the suspension downstream from the grid as particles are released from the electrodes, after removal of the applied voltage, and pass through a light beam directed into the cell via an optical fibre.

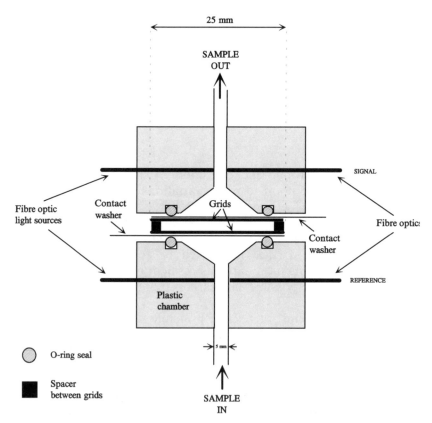

Figure 5.12   Dielectrophoresis cell with grid electrodes that allow sample flow perpendicular to the plane of the electrodes

In the demonstrated prototype two grids are pressed against a thin, insulating spacer and held together in a nylon support by screws. Although successful as a laboratory demonstrator, this simple structure lacks the robustness and flexibility needed for reliable on-line dielectrophoretic sensing. Electrode cells comprising parallel pairs of grids in principle can be fabricated as an integrated structure by using the adaptation of VLSI technology commonly known as micromachining. Figure 5.13 shows an SEM microphotograph of a micromachined electrode of a prototype grid formed by nickel electroplating.[2] The objective is to develop a self-supporting, integrated grid comprising the two electrodes separated by a very thin insulating layer. To achieve the desired robustness the electrode shown is ~25 μm thick. The pores have a diameter of 50 μm, to allow a reasonable flow and to give a sufficient edge length as the collecting surface. The electrode is formed by electroplating a seed layer through a patterned polyimide layer that acts as the mould for the shaped metal structure [19]. Other layers, for example insulators and other grid electrodes, can be deposited on top of the lower, supporting layer and patterned simultaneously.

Other innovative electrode designs include a three dimensional electric field cage formed by the accurate alignment and fixing of two quadrupole structures of the form shown in Figure 5.7 [20]. Also spiral electrode structures that manipulate microbiological cells by the motion induced by a travelling electric field have been demonstrated [6, 21]. Multi-level metallisation schemes have been developed in which the

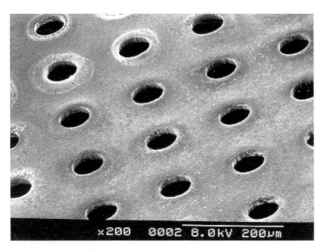

Figure 5.13 *SEM microphotograph of a self-supporting grid electrode formed by nickel electroplating through a polyimide micro-mould. The nickel film is ~25 μm thick*

[2] REYNOLDS, D. N., and ALLSOPP, D. W. E.: unpublished

lower level metal forms the interconnect between a longitudinal array of electrodes that are addressed in sequence [22]. The sequencing of the electrode bias creates a travelling electric field which often induces characteristic particle motion by a process related to negative dielectrophoresis.

In summary, a dielectrophoretic sensor or analyser requires, at its heart, a set of metallic electrodes by which the polarising electric field is applied. VLSI circuit fabrication techniques or micromachining methods can be used to construct robust electrode cells at potentially low cost, whilst the design and configuration of the electrodes is largely determined by the application. The interfacing with the instrumentation needed for monitoring the dielectrophoretic processes and other electronics is another important consideration in the design of the electrode cell.

5.5 Microbiological applications of dielectrophoresis

Microbiology has long been recognised as an area in which to exploit the capability of dielectrophoresis. Microbiological cells are surrounded by a series of layers that make up a membrane containing mobile ions and other free charges, rendering the whole cell susceptible to polarisation by an applied electric field [1, 2]. Small differences in the structure of the membrane, for example in its ionic content, can give rise to changes in its conductivity and dielectric properties. As a result microbiological cells can display a frequency dependent, possibly characteristic response to a dielectrophoretic force when contributions to the overall dielectrophoretic response from the medium containing the particles are factored out. Further, the surface and subsurface layers of a cell can be particularly sensitive to changes in the chemical environment, for example to the introduction to the suspension of contamination or a drug that affects their viability or vitality. These are the features, combined with the ability to accrete polarised but otherwise neutral particles, that make dielectrophoresis a promising technique for meeting the requirements of pre-concentration, separation and even quantification in microbiological analysis, with the added advantage that such sensing can be on-line.

The potential and early applications of dielectrophoresis to biological analysis have been described by Pohl [1] and by Pethig [2]. Rather than summarising these comprehensive works (interested readers are referred to these), this section concentrates instead on reviewing a few more recent studies of micro-organisms by dielectrophoresis in which sensor systems of the type outlined in the previous section have been used.

Dielectrophoretic collection of colloidal yeast from aqueous suspensions under both positive and negative forces has been widely investigated [1, 3, 4, 7]. An example of the collection of yeast under a positive dieletrophoretic force has already been shown in Figure 5.4. In a potentially powerful development the separation of mixed populations of cells by simultaneous positive and negative dielectrophoretic forces has been demonstrated using specially designed electrodes [4]. An application of this technique to the separation of sheep erythrocytes (blood cells) and bacteria is shown in Figure 5.14.

The bacteria have collected at the electrode edges under positive dielectrophoretic force, whilst the blood cells collect preferentially in the inter-electrode region and on the electrodes under negative dielectrophoretic force. The ability to separate related but distinguishable microbiological particles using combined positive and negative dielectrophoresis has wide potential applications. In the example reviewed here, differentiating between viable and non-viable yeast cells is of likely significance for the brewing and related industries, as fermentation rates will be determined by concentration of viable yeast in a given mash or process.

Simultaneous positive and negative dielectrophoresis has also been used to remove from blood human leukaemia cells [23] and breast cancer cells [24].

Another area of microbiology where dielectrophoresis has been applied to address problems of topical interest is that of the purity of drinking water supplies. Waterborne microbes can present a serious health hazard and the related issues of their detection, safe concentrations and the use of disinfectants will be controlled by stringent EC

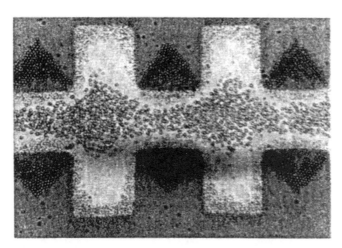

Figure 5.14   *Separation of a previously mixed suspension of sheep erythrocytes*
                and bacteria by simultaneous positive and negative dielectrophoretic
                forces that result from specially designed electrodes. Data
                reproduced by permission from Wang et al. [4]

legislation. For example, the protozoan parasite *Cryptosporidium parvum* is known to infect human beings, causing severe gastroenteritis, even mortality, in immunocompromised patients, and legislation will probably require drinking water to be tested for its presence in a biologically viable form. Equally, new legislation will place upper limits on the concentrations of disinfectants such as chlorine and ozone present in potable water.

Dielectrophoresis has been found to be an effective method for not only the rapid detection of *C. parvum* in its oocyst stage in water, but also for differentiating between untreated, autoclaved (heat treated) and ozone treated oocysts [11, 18, 21, 25, 26]. Figure 5.15 compares the frequency dependence of the collection efficiency under positive dielectrophoretic force of untreated and ozone treated *C. parvum* oocysts. The effects of two different concentrations of ozone are shown. The spectra of dielectrophoretic collection efficiency were obtained using co-planar aluminium electrodes of 5 μm separation with an applied ac voltage of 12 V peak-to-peak at frequencies in the range 1 kHz to 10 MHz [11]. Sharp differences in the collection efficiency at frequencies of ~100 kHz are observed between the treated and untreated samples and even for the two ozone dosages. The differences in dielectrophoretic collection efficiency are thought to be caused by the ozone rupturing the oocyst membrane and disrupting the transport of ions in the membrane [25].

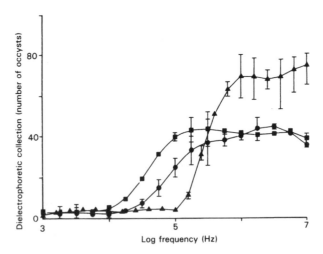

Figure 5.15   Comparison of the frequency dependence of the dielectrophoretic collection efficiency of untreated and ozone treated Cryptosporidium parvum *oocysts: triangles, untreated oocysts; circles, 1.15 mg/l ozone dosage; squares, 3.3 mg/l ozone dosage (from Reference 10)*

Whilst this result underlines the potential of the technique for discriminatory on-line environmental sensing, its full potential will require electrode structures capable of allowing a greater throughput of liquid. Here the performance of the parallel grid electrodes described in the previous section is considered. Simple grid electrodes made from membrane pre-filters have been shown to separate untreated *C. parvum* oocysts from ozone-treated oocysts [18]. Again sufficient discrimination between the dielectrophoretic responses of the treated and untreated *C. parvum* oocysts was obtained; however, their relative collection efficiencies differed from those observed using the coplanar electrodes. As yet this result is not understood, but is thought to be due to the different balance between the counteracting forces derived from the hydrostatic pressure and the related flow rate of the medium.

5.6 Summary

The nascent and growing field of microbiological sensing by dielectrophoresis has been reviewed, with emphasis on the engineering aspects of instrumentation and transducer design. Dielectrophoresis involves the polarisation of a neutral particle by an alternating electric field and the subsequent motion of the polarised particle in the field. The effect results from the microscopic differences in the Coulombic forces experienced by the opposite poles of the induced dipole when the driving field is nonuniform. It is an electromagnetic phenomenon, but one that needs a detailed understanding of the dielectric response of particulate matter. As such, studies of the dielectrophoretic response of particles can be regarded as a cross-disciplinary science where, for the area of application described here, detailed knowledge of the make-up of microbiological particles and of electrical science is required if the full potential of this technique is to be realised.

The basic theory of dielectrophoresis was reviewed following the approach first used by Pohl [1]. Next the principles of sensing of particles by dielectrophoretic response were discussed and the need for a sensing system was established. The requirements of such sensor systems and their formation are determined by the type and size of the particles of interest and, to some extent, on the attribute of dielectrophoresis being exploited. Some practical sensor systems were reviewed in Section 5.4. In Section 5.5 recent examples of the application of dielectrophoretic sensing systems to detecting and characterising microbiological particles were presented. The examples given and the review of the electronic engineering issues regarding both the design and fabrication of the electrodes needed for dielectrophoresis through to the creation of complete sensor systems were intended to give readers a forward look at the technique of dielectrophoresis, its application and its potential, rather than to give an exhaustive review of the subject. In this respect it is hoped that, by looking forward, new ideas and new applications for dielectrophoresis can be devised.

5.7 References

1 POHL, H.: 'Dielectrophoresis' (CUP, Cambridge, 1978)
2 PETHIG, R.: 'Dielectric and electronic properties of biological materials' (Wiley, Chichester, 1979)
3 PETHIG, R., HUANG, Y., WANG, X.-B., and BURT, J. P. H.: 'Positive and negative dielectrophoretic collection of colloidal particles using interdigitated castellated microelectrodes', *Journal of Physics D*, 1992, **24**, pp. 881–888
4 WANG, X.-B., HUANG, Y., BURT, J. P. H., MARKX, G. H., and PETHIG, R.: 'Selective dielectrophoretic confinement of bioparticles in potential energy wells', *Journal of Physics D*, 1993, **26**, pp. 1278–1285
5 FUHR, G., ARNOLD, W. M., HAGEDORN, R., MÜLLER, T., BENECKE, W., WAGNER, B., and ZIMMERMANN, U.: 'Levitation, holding and rotation of cells within traps made by high frequency fields', *Biochimica et Biophysica Acta*, 1992, **1108**, pp. 215–223
6 WANG, X.-B., HUANG, Y., WANG, X., BECKER, F. F., and GAS-COYNE, P. R. C.: 'Dielectrophoretic manipulation of cells with spiral electrodes', *Biophysical Journal*, 1997, **72**, pp. 1887–1899
7 HAWKES, J. J., ARCHER, G. P., and BETTS, W. B.: 'A dielectrophoretic spectrometer for characterising micro-organisms and other particles', *Microbios*, 1993, **73**, pp. 81–86
8 HUANG, Y., and PETHIG, R.: 'Electrode design for negative dielectrophoresis', *Measurement Science & Technology*, 1991, **2**, pp. 1142–1146
9 MÜLLER, T., GERARDINO, A. M., SCHNELLE, T., SHIRLEY, S. G., FEHR, G., DE GASPERIS, G., LEONI R., and BORDONI, F.: 'High-frequency electric-field trap for micron and submicron particles', *Il Nuovo Cimento*, 1995, **17D**, pp. 425–432
10 PETHIG, R., 'Application of ac electric fields to the manipulation and characterisation of cells', in KARUBE, I. (Ed.): 'Automation in biotechnology' (Elsevier, 1991), pp. 159–185
11 ARCHER, G. P., BETTS, W. B., and HAIGH, T.: 'Rapid differentiation of untreated, autoclaved and ozone-treated *Cryptosporidium parvum* oocysts using dielectrophoresis', *Microbios*, 1993, **73**, pp. 165–172
12 KALER, K. V. I. S., and JONES, T. B.: 'Dielectrophoretic spectra of single cells determined by feedback-controlled levitation', *Biophysical Journal*, 1990, **57**, pp. 173–182
13 GREEN, N. G., and MORGAN, H.: 'Dielectrophoretic separation of nanoparticles', *Journal of Physics D*, 1997, **30**, pp. L41–L44
14 MORGAN, H., and GREEN, N. G.: 'Dielectrophoretic manipulation of rod shaped viral particles', *Journal of Electrostatics*, 1997, **42**, pp. 279–293
15 MILNER, K., BROWN, A. P., ALLSOPP, D. W. E., and BETTS, W. B.:

'Dielectrophoretic collection of bacteria using differential impedance measurements', *Electronics Letters*, 1998, **34**, pp. 66–67

16 MILNER, K. R., BROWN, A. P., BETTS, W. B., GOODALL, D. M., and ALLSOPP, D. W. E.: 'Analysis of biological particles using dielectrophoresis and impedance measurement'. Proceedings 19th Rocky Mountain Bioengineering Symposium, Colorado, USA, 1998

17 ARCHER, G. P., RENDER, M. C., BETTS, W. B., and SANCHO, M.: 'Dielectrophoretic collection of micro-organisms using grid electrodes', *Microbios*, 1993, **76**, pp. 237–244

18 ARCHER, G. P., QUINN, C. M., BETTS, W. B., SANCHO, M., MARTINEZ, G., LLAMAS, M., and O'NEILL, J. G.: 'An electrical filter for the selective concentration of cryposporidium parvum oocysts from water'. Proceedings of Royal Society Chemical Conference on *Protozoan parasites and water*, York, September 1994 (Royal Society of Chemistry, Cambridge, 1995), pp. 146–151

19 FRAZIER, A. B., and ALLEN, M. G.: 'Metallic microstructures fabricated using photosensitive polyimide electroplating molds', *Journal of Microelectromechanical Systems*, 1993, **2**, pp. 87–94

20 SCHELLE, T., HAGEDORN, R., FUHR, G., FIEDLER, S., and MÜLLER, T.; 'Three dimensional electric field traps for manipulation of cells – calculation and experimental verification', *Biochimica et Biophysica Acta*, 1993, **1157**, pp. 127–140

21 GOATER, A. D., BURT, J. P. H., and PETHIG, P.: 'A combined travelling wave dielectrophoretic and electrorotation device: applied to the concentration and viability determination of cryptospiridium', *Journal of Physics D*, 1997, **30**, pp. L65–L69

22 MORGAN, H., GREEN, N. G., HUGHES, M. P., MONAGHAN, W., and TAN, T. C.: 'Large-area travelling wave dielectrophoresis particle separator', *Journal of Micromechanics and Microengineering*, 1997, **7**, pp. 65–70

23 BECKER, F. F., WANG, X.-B., HUANG, Y., PETHIG, R., VYKOUKAL, J., and GASCOYNE, P. R. C.: 'The removal of human leukaemia cells from blood using interdigitated microelectrodes', *Journal of Physics D*, 1994, **27**, pp. 2659–2662

24 GASCOYNE, P. R. C., WANG, X.-B., HUANG, Y., and BECKER, F. F.: 'Dielctrophoretic separation of cancer cells from blood', *IEEE Transactions on Industrial Applications*, 1997, **33**, pp. 670–678

25 QUINN, C. M., ARCHER, G. P., BETTS, W. B., and O'NEILL, J. G.: 'Dose dependent dielectrophoretic response of cryptospiridium oocysts treated with ozone', *Letters in Applied Microbiology*, 1996, **22**, pp. 224–228

26 ARCHER, G. P., QUINN, C. M., BETTS, W. B., ALLSOPP, D. W. E., and O'NEILL, J. G.: 'Physical separation of untreated and ozone-treated *Cryptosporidium parvum* oocysts using non-uniform electric fields'. Proceedings of the Royal Society of Chemistry Conference on *Protozoan parasites and water*, York, September 1994 (Royal Society of Chemistry, Cambridge, 1995), pp. 143–145

Chapter 6

Electrically conducting polymers for sensing volatile chemicals

K. C. Persaud and Siswoyo

6.1 Introduction

A large number of organic, inorganic or organometallic compounds can be polymerised. Their physical, chemical and electrical properties depend on complex sets of interacting factors, that include molecular structure, molecular size and cross-link density. Of the many types of polymeric materials, electrically conducting organic polymers have attracted much interest because their unusual properties lend themselves to many novel applications. Previously, organic polymers were considered good insulators. However, this concept was radically changed when it was shown that the conductivity of polyacetylene can be increased via doping by 13 orders of magnitude, up to about 10^4 S cm^{-1}. The class of polymers known as electrically conducting polymers was born [1].

Conducting polymers are conjugated organic materials, and these differ from redox polymers in that the polymer backbone is itself conducting (see Figure 6.1). The conductivity of the materials can be modulated from the non-conducting form to highly conducting form by the addition of dopant ions into the polymer matrix. This can be accomplished electrochemically or via purely chemical pathways.

Although there is a large body of research on polyacetylenes, the instability of many of these materials in air, until recently, limited the number of practical applications. Of the many conducting polymers investigated, it can be said that polypyrrole is currently the most popular. Chemical methods for the oxidative polymerisation of pyrrole had been known for many years, but when Diaz *et al.* [2] produced the first free-standing films of electrochemically deposited polypyrrole, poly-N-methylpyrrole and poly-N-phenylpyrrole on platinum electrodes in an acetonitrile/ Et$_4$NBF$_4$ solution, a new era in the development of novel materials arose.

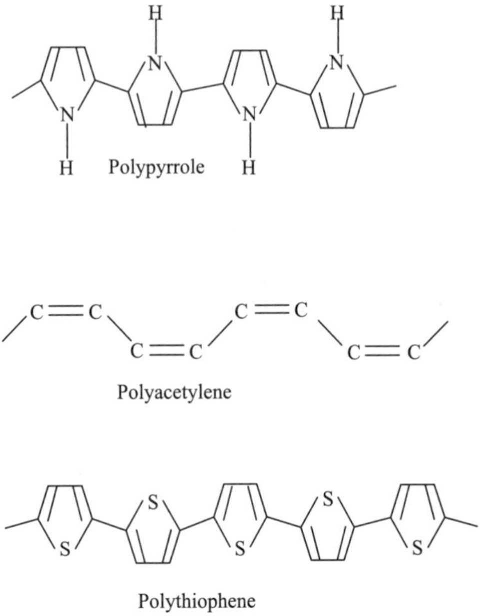

Polypyrrole

Polyacetylene

Polythiophene

Figure 6.1   Some repeating structures of organic conducting polymers

All of the compounds which have successfully produced conducting films have several common characteristics. They are aromatic and can be oxidised at relatively low anodic potentials. The aromatic compound undergoes electrophilic substitution reactions where the aromatic structure is maintained.

6.1.1 Synthesis of conducting polymers

The major techniques used to synthesise conducting polymers are of four types:

- pyrolysis
- Ziegler–Natta catalysis
- electrochemical synthesis
- condensation polymerisation.

6.1.1.1 Pyrolysis

Preparation of a conducting polymer using this technique consists of eliminating heteroatoms (such as halogens, oxygen and nitrogen) from the polymer by heating it to form an extended aromatic structure eventually approaching that of graphite. Graphite behaves as a prototypical synthetic metal when treated with certain dopants. Pyrolysis probably increases charge carrier mobilities by producing extended conjugation and increases the number of charge carriers by forming free radicals, which can act as donors to form hole carriers (cations) or as acceptors to form electron carriers (anions). The products of polymer pyrolysis can be powder, film or fibre, depending on the form and nature of the starting polymer and the pyrolysis conditions.

6.1.1.2 Ziegler–Natta catalysis

Of the catalysts available, the Ziegler–Natta catalyst [Ti(O-*n*-C$_4$H$_9$)$_4$–Al(C$_2$H$_5$)$_3$] system in its various modifications is by far the most used even today for synthesis of polyacetylene. It yields free standing films as well as foams, partially oriented stretched films, composites and powders depending on the experimental procedure used. All these generated materials, however, have been insoluble and infusible. Shirakawa's method of making silvery polycrystalline films of (CH=CH)$_n$ using a Ziegler–Natta catalyst is the most widely used and best characterised process [3]. Doping polyacetylene can increase its conductivity by 12 orders of magnitude and alter its electrical, optical and magnetic properties. The doping can be either p-type (oxidative), with bromine, iodine, AsF$_5$, HClO$_4$ or Ag(ClO$_4$), or n-type (reductive), with lithium, sodium or potassium.

6.1.1.3 Electrochemical synthesis

A good example of a conducting polymer prepared by the technique of electrochemical synthesis is polypyrrole. Films prepared by the stoichiometric electropolymerisation of pyrrole in an electrolyte solution containing tetraethyl ammonium tetrafluoroborate have conductivities in the range of 40 to 100 S cm^{-1}. A combination of chemical and electrochemical analysis indicates that the structure of polypyrrole prepared in tetraethyl ammonium tetrafluoroborate electrolyte is as shown in Figure 6.2.

A typical polymerisation cell consists of a single compartment electrochemical cell containing two electrodes. The anode may be an

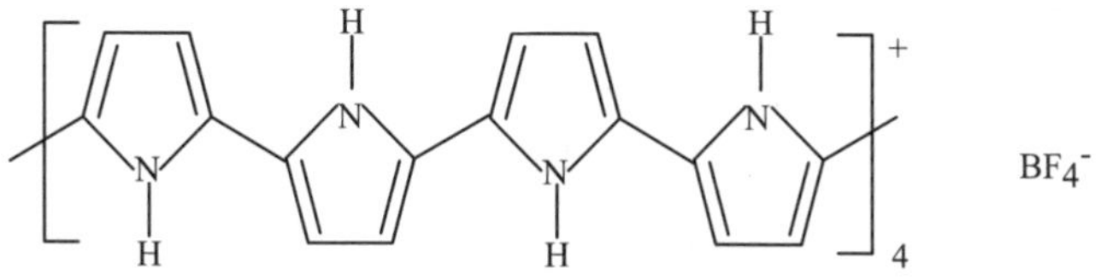

Figure 6.2   Polypyrrole tetrafluoroborate

evaporated platinum electrode and the cathode can consist of gold wire tightly wrapped around a glass microscope slide. A plane-parallel cell arrangement is necessary to ensure uniform thickness over the film. Typically, the cell may contain 0.1 M Et_4NBF_4 and 0.006–0.1 M pyrrole in 99% aqueous CH_3CN, but several variations in the type of counterion and concentration of monomer are permitted. For electropolymerisation, the cell is normally under galvanostatic control, although potentiostatic methods can also be used. The final current density is usually set at 1 mA cm^{-2}, although values ranging from 0.5 to 1.5 mA cm^{-2} may be used. Once the final current density is reached, little further change in the cell potential is noted, indicating that the film being deposited continues to conduct. The platinum electrode gradually darkens as the polypyrrole layer thickens. Films of up to 50 μm thick are prepared in this manner.

6.1.1.4 Condensation polymerisation

Condensation polymerisation reactions are widely used in the field of polymerisation. The polymers that have been successfully synthesised include poly(p-phenylene), poly(p-phenylene sulphide), polyquinolines, polyquinoxalines and pyrrones.

Synthesis of poly(p-phenylene) can be done by the reaction shown in Figure 6.3, but this only yields oligomeric material and even this is insoluble. However, radical polymerisation of 5,6–dihydroxycyclohexa-1,3–diene (Figure 6.4) leads to a soluble precursor polymer that can be processed prior to the final thermal conversion into poly(p-phenylene). The material is an insulator in the pure state but can be both n- and p-doped using methods similar to those used for polyacetylene.

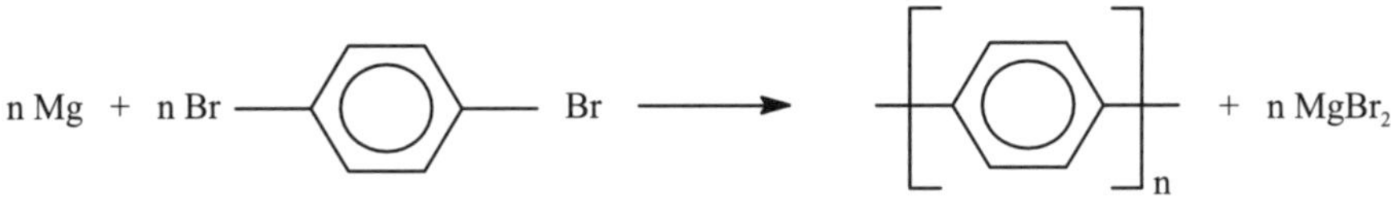

Figure 6.3.   *Polycondensation reaction for formation of poly(phenylene)*

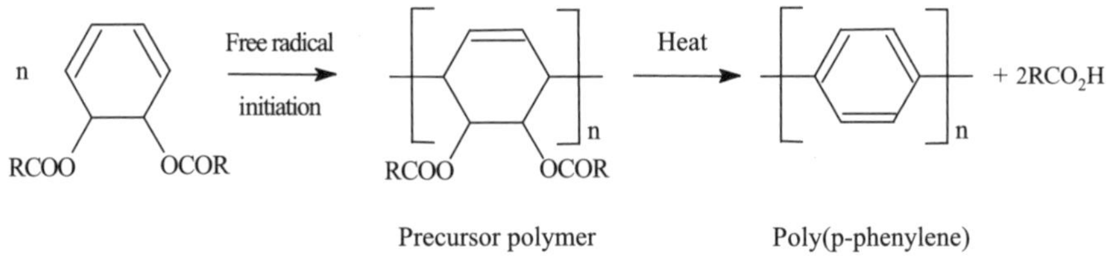

Figure 6.4   *Synthesis of poly(phenylene)*

6.2 Conduction models for conducting polymers

The electrical conduction properties of elemental semiconductors can be controlled by incorporating very small quantities of foreign atoms into the host semiconductor lattice. The host can be made n-type or p-type depending on the nature of the added dopant atoms. New dopant energy levels are introduced into the band-gap and conduction is facilitated.

A similar concept of doping has been applied to conjugated polymers. These materials may also be doped, and the conductivity level obtained depends on the doping level. However, the doping mechanism differs considerably from that observed for semiconductors. The electrical conductivity depends on a number of fundamental parameters, such as the number density of mobile charge carriers n, the carrier charge q and the carrier mobility μ. The relationship between these parameters and the conductivity σ is expressed by the general relationship:

$$\sigma = nq\mu \tag{1}$$

Conduction in solids can be explained in terms of band theory (Figure 6.5), which postulates that when atoms or molecules are aggregated in the solid state, the outer atomic orbitals containing the valence electrons are split into bonding and antibonding orbitals, and mix to form two series of closely spaced energy levels, usually called the valence band and conduction band, respectively. If the valence band is only partly filled by the available electrons, or if the two bands overlap so that no energy gap exists between them, then application of a potential will raise some of the electrons into empty levels where they will be free to move throughout the solid, thereby producing a current. This is the description of a conductor. On the other hand, if the valence band is full and is separated

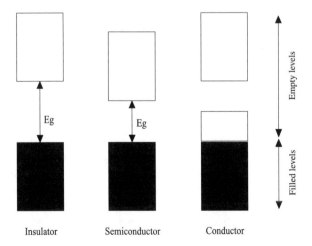

Figure 6.5   Schematic diagram showing the principles of band theory

from the empty conduction band by an energy gap, then there cannot be net flow of electrons under the influence of an external field unless electrons are elevated into the empty band, and this will require a considerably expenditure of energy. Such materials are either semiconductors or insulators, depending on how large the energy gap may be.

Electronic conduction in polymers cannot be totally explained by using band theory, because the atoms in conducting polymer are covalently bonded to one another, forming a polymeric chain that experiences weak intermolecular interaction. Thus conduction will require electron movement, not only along chains but also from one chain to another.

Polymers have the electronic profiles of either insulator or semiconductor; thus the band-gap in a fully saturated chain such as polyethylene is 5 eV and decreases to about 1.5 eV in the conjugated system polyacetylene. The respective intrinsic conductivities are $\sim 10^{-17}$ S cm^{-1} and $\sim 10^{-8}$ S cm^{-1}.

From the viewpoint of structure, polyacetylene is perhaps the simplest type of conducting polymer. Two of the three p orbitals of carbon atoms in polyacetylene are in the form of sp^2 hybrid orbitals, giving rise to the σ bonding of the polymer, with the third entering into a covalent bond with the hydrogen atom s orbitals. The third p orbital forms an extended π system along the carbon chain. In a polyconjugated system the π orbitals are assumed to overlap, and form a valence and a conduction band as predicted by band theory. If all the bond lengths were equal, then the bands would overlap and the polymer would behave like a quasi-one dimensional metal having good conductive properties. However, experimental evidence does not substantiate this and reference to the physics of a monoatomic one dimensional metal, with a half-filled conduction band, has shown that this is an unstable system and will undergo lattice distortion by alternate compression and extension of the chain.

The trans-structure of polyacetylene is unique as it has a two-fold degenerate ground state in which sections A and B are mirror images and the single and double bonds can be interchanged without changing the energy. Thus if the cis structure begins to isomerise to the trans-geometry from different locations in a single chain, an A sequence may form and eventually meet a B sequence, as shown in Figure 6.6, but in doing so, a free radical is produced. This is a relatively stable entity and the resulting defect in the chain is called a neutral soliton. The electron has an unpaired spin and is located in a non-bonding state in the energy gap, mid-way between the two bands. It is the presence of these neutral solitons which gives trans-polyacetylene the characteristics of a semiconductor with an intrinsic conductivity of about 10^{-7} to 10^{-8} S cm^{-1} [4].

On the basis of the soliton model, many unusual properties have been predicted for all trans-polyacetylene, including a mechanism by which conductivity can be explained. However, other polymers, such as

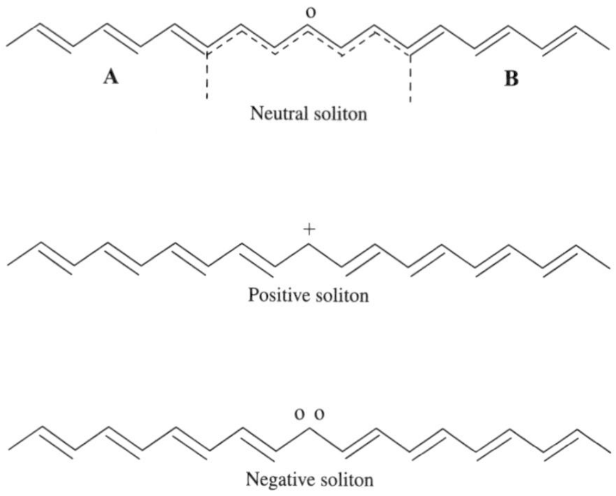

Figure 6.6   Schematic representation of solitons in polyacetylene

polyparaphenylene, do not possess degenerate ground states and are not expected to accommodate solitons; nevertheless, doped polyparaphenylene displays transport properties which are very similar to those of doped polyacetylene [5] (Figure 6.7). In this case the concept of soliton transport cannot be used, since if two regions separated by topological defect are not energetically degenerate, then formation of single solitons is energetically unfavourable.

From Figure 6.8, it can be seen that unpaired electrons are generated where the benzenoid and quininoid structures meet. It is well-established in solid state physics that if a charge carrier is localised and trapped, it tends to polarise the local environment, which then relaxes into a new equilibrium position. This local deformed section of the polymer chain and the charge carrier are then termed a polaron. Unlike the soliton, the polaron cannot move without first overcoming an energy barrier, so movement is by a hopping process. In polyphenylene the solitons are trapped by the changes in polymer structure because of the differences in energy and so a polaron is created, which is an isolated charge carrier. A pair of these charges is called a bipolaron.

At high doping levels bipolarons interact to form bipolaron bands within the energy gap. Hence a general picture for polymers with a non-degenerate ground state is as follows. The neutral polymer has full valence and empty conduction bands, separated by a band-gap. Electrochemical doping removes one electron, resulting in the generation of a polaron level located at mid-gap. Further oxidation results in the removal of a second electron to generate a bipolaron. Still further results generate bipolaron energy bands in the band-gap. Electronic conductivity is rationalised in terms of bipolaron hopping [6].

As well as polyphenylene, polypyrrole and other related conjugated

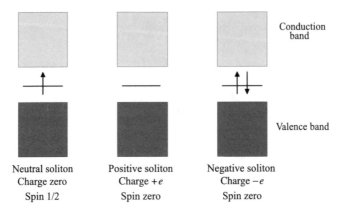

Neutral soliton	Positive soliton	Negative soliton
Charge zero	Charge +*e*	Charge −*e*
Spin 1/2	Spin zero	Spin zero

Figure 6.7   Schematic representation of neutral, positive and negative solitons

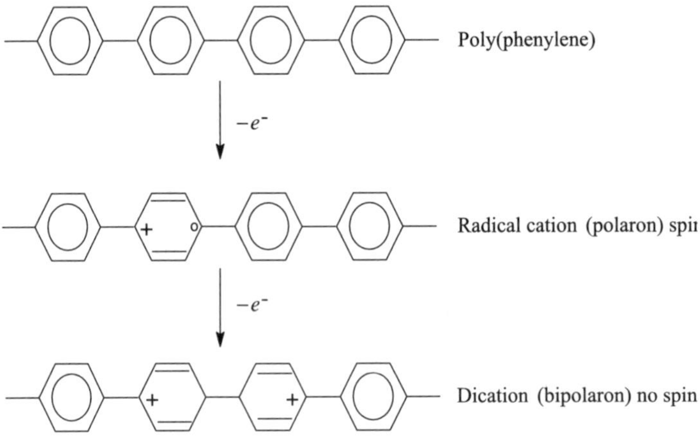

Poly(phenylene)

−*e*⁻

Radical cation (polaron) spir

−*e*⁻

Dication (bipolaron) no spin

Figure 6.8   Generation of polaron and bipolaron defects in polyphenylene

polymers are thought to posses a charge-carrying mechanism involving bipolarons (Figure 6.9). Bipolarons are formed by the interaction of two polarons (Figure 6.9a), which in turn are formed by interaction of a neutral and a charged soliton. A positive bipolaron is therefore a delocalised dication, the presence of which can induce large changes in conductivity. The incorporation of anions in the polymer structure stabilises the polarons and bipolarons, although it is thought that the incorporation of further dopant can induce polaron and bipolaron formation leading to a change in conductivity.

Bredas *et al.* [6] reported an *ab initio* restricted Hartree–Fock (RHF) study of the electronic and geometric structures of undoped and highly doped quaterpyrrole as a model chain for polypyrrole. By using the valence-effective-Hamiltonian (VEH) method they also predicted a

(a)

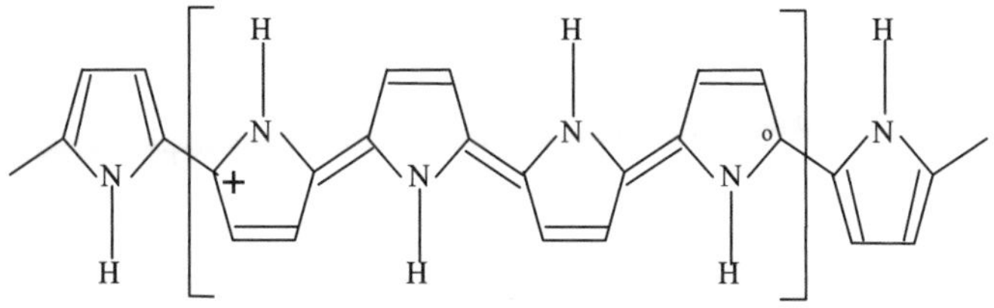

(b)

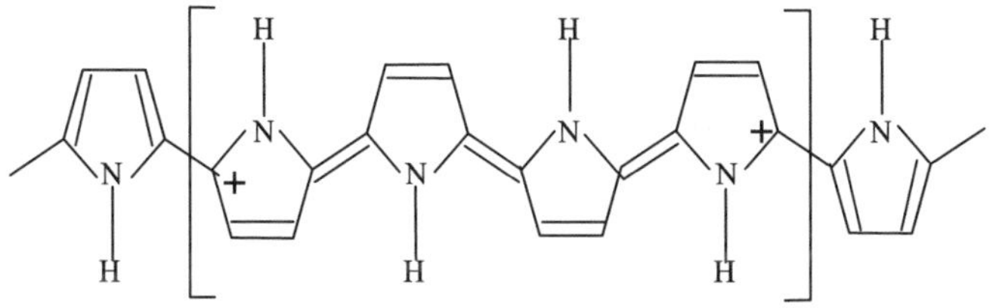

Figure 6.9   Polaron (a) and bipolaron (b) formation in polypyrrole

picture of the band structure evolution upon high doping of Ppy (Figure 6.10). For an undoped Ppy chain, a band-gap of 4.0 eV was obtained. By analogy with polyphenylene, at low doping levels it is expected that polarons rather than bipolarons are present on the chain. At intermediate doping levels, the presence of bipolarons on the chains would provoke the appearance of bipolaron states in the gap located 0.46 eV above the valence band (VB) edge and 0.89 eV below the conduction band (CB) edge (Figure 6.10b). As the doping level increases, the bipolaron states overlap. This process results in the formation of bipolaron bands in the gap. This is what occurs for a 33% doping level per monomer (Figure 6.10c). If still higher doping levels could be achieved, bipolaron bands eventually merge with the CB and VB. Conductivity with spin occurs, as a result of the partially filled character of the CB or VB in the case of n- or p-type doping, respectively.

There is experimental evidence from electron spin resonance measurement (ESR) [now known as EPR (electron paramagnetic resonance)] [7] and optical spectroscopy [8] that supports the predictions that spinless bipolarons are connected with the high conductivity of polypyrrole. EPR measurements were made on a polypyrrole film exposed to various pressures of oxygen. Subsequently the conductivity was seen to level off whilst the EPR signal fell to zero. The initial rise in conductivity was explained in terms of the formation of polarons which have both charge and spin. The fall in EPR signal at higher levels of doping results from the formation of the more energetically favourable bipolarons, which also have charge but are spinless.

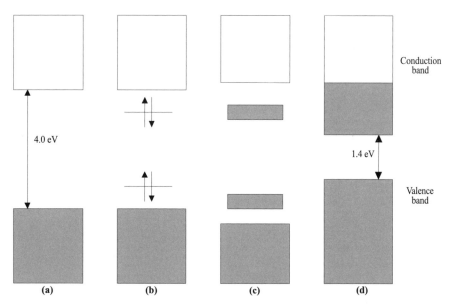

Figure 6.10   VEH band–band structure evolution upon (sodium) doping of polypyrrole: (a) undoped; (b) intermediate doping levels: non-interacting bipolarons present on the chain; (c) 33% doping level per monomer; (d) 100% doping level per monomer

6.3   Conducting polymers as sensors for volatile chemicals

From the above discussion, it can be realised that the electrical properties of conducting polymers can be readily modulated by the introduction of dopant ions. However, the adsorption/desorption of volatile chemicals on to conducting polymers can also lead to modulation of conductivity. This has been exploited for the use of these materials as gas and vapour sensors. A reversible change of dc resistance on exposure of many conducting polymers to a gas or vapour sample is observed. Miasik's group [9] have shown that gas sensing devices can be fabricated using electronically conducting polymers for the ambient temperature detection of several industrially important gases. In particular, the resistance of thin films of polypyrrole increased in the presence of 0.1% ammonia in air and decreased in the presence of 0.1% nitrogen dioxide and 0.1% hydrogen sulphide.

The resistance of the polypyrrole film changes in the presence of methanol vapour, the response being rapid and reversible at room temperature. Furthermore, the time course and concentration dependence of the resistance changes are consistent with a model in which the methanol interacts with sites within or on the surface of the film.

These polymers have a number of distinct advantages from the point of view of gas sensing, available in a wide variety including polypyrroles, polythiophenes, polyindoles and polyanilines, readily grown by electrochemical polymerisation of the monomer and can be operated at room temperature. One of the disadvantages of these materials is their lack of specificity; they respond to a wide range of different gases and vapour. However, these materials can be used in an intelligent gas sensor if arrays are made of different types of polymers [10, 11].

The response mechanism of the conducting polymer polypyrrole to a selection of gases and vapours was investigated using two techniques: measurement of resistance change and mass change with the objective of characterising responses for incorporation into a sensor array [12]. Its conductivity response has a dual nature. Firstly, there is the widely accepted view of nucleophilic gases causing increases in resistance with electrophilic gases having the opposite effect. Secondly, certain vapours have a solvent type action on the polymer, causing it to swell. Changes in dimension of the polymer are accompanied by conductivity change; hence changes of apolar analytes also change the polymer conductivity, possibly affecting the activation energy for electron hopping between chains, and thus the conductivity decreases.

Much interest exists in the use of such sensor materials for discriminating between odours, and the potential industrial applications of these sensors in the area of quality control in the food and beverage industries, malodours in the environment and clinical diagnostics, are huge.

6.3.1 Assessment of odour quality and concentration

In order to understand how the sensors can be applied to odour measurement, it is worth investigating how existing methods of odour measurement actually operate.

An odour is normally composed of many chemical species mixed together in different proportions and which the receptors in the nose perceive all at once, without separation into individual chemical species. An individual will normally describe an odour in terms of previous experience, but this needs to be standardised if an odour description is required for an industrial application. The complex mixtures of chemicals found in 'normal odours' present us with a difficult classification problem. The character of an odour is reported using 'odour descriptors' such as 'fruity', 'floral', 'musky', 'burnt', 'smoky', 'rancid', etc. There are no consistent ways of standardising the descriptors of odours, and many individuals and companies have compiled lists of descriptors that may be used to train persons who make up a human odour panel. Since the odour panel may be specialised to assess only certain products, or odour quality problems, the lists of relevant descriptors are often expanded or contracted as required. In order to assess odour quality, a specialised vocabulary needs to be learnt, with each descriptor associated with specific notes, or nuance in a particular odour. Other aspects of odour measurement are also important. The hedonic tone of an odour is a measure of the pleasantness or unpleasantness of an odour sample. This hedonic tone is independent of the character of the odour, and is often measured on a psychophysical scale, that may typically range from −10 to +10, where 0 represents neutral, −10 highly unpleasant and +10 very pleasant. The process of assigning a hedonic tone value is highly subjective, and relies on an internal referencing scale of the panellist, which is often dependent on the degree of training and experience of the panellist.

Determination of how much of an odour is present is often carried out by olfactometry. Odour samples are evaluated using an instrument called an olfactometer with a human odour panel, typically consisting of four to eight persons. Both European and American Standards for olfactometry exist, and odours are evaluated in accordance with defined standard practices. The protocols followed in Europe are described in the CEN/TC264 (Comitée Européen de Normalisation) draft standard. The technique uses a dynamic dilution olfactometer, which has a forced choice method sample presentation to an odour panel. Odour panels consist of individuals (panellists) that are selected and trained. Typically only people with a consistent personal threshold for the reference standard n-butanol detection of between 20 ppb and 80 ppb are used, and these assessors are continually checked for their detection threshold. An odour panel is conducted using a panel of individuals (panellists) and an olfactometer. Many olfactometer designs exist, but typically sample odour is delivered at a fixed rate from a Tedlar™ gas-sampling bag to the olfactometer. A filtered air system delivers non-odorous dilution air to the olfactometer. The trained odour panellist sniffs the diluted odour sample as it is discharged from one of two or three sniffing ports. The panellist sniffs all sniffing ports and must select one that is different from the others. This is a forced choice approach. The panellist decides if the selection was a 'guess', 'detection' or 'recognition'. The panellist is then presented with progressively higher concentrations, each time increasing by a factor of two. This statistical approach is called an 'ascending concentration series'. This method is a rapid means of determining sensory thresholds of any substance usually in an air medium. The threshold may be characterised as being either (a) only detection (awareness) that a very small amount of added substance is present but not necessarily recognisable, or (b) recognition of the nature of the added substance. It is recognised that the degree of training received by a panel with a particular substance may have a profound influence on the threshold obtained with that substance, and that thresholds determined by using one physical method of presentation are not necessarily equivalent to values obtained by another method.

The numbers of binary dilutions of the odour sample to human threshold give objective numerical measures of 'odour units' present in the sample. While these are dimensionless units, dimensions of odour unit per unit volume are commonly applied. Odour concentration is commonly expressed as European odour units per m^3 of air (O_E m^{-3}). Odour concentrations have been previously reported in a variety of ways that are difficult to compare. These include detection threshold, recognition threshold, dilution to threshold, effective dose at 50 percentile, dilution ratio, odour units, odour dilution units and best estimated threshold.

For odour intensity measurements, the general objective is to reference the odour intensity rather than other odour properties of a sample. This is done by a comparison of the odour intensity of the sample to the odour intensities of a series of concentrations of the reference odorant, which is typically 1-butanol. The odour intensity is the relative strength of the odour above the threshold. The odour intensity of a sample is estimated by a comparison of the odour intensity of the sample to the odour intensity of a series of concentrations of butanol increasing in concentration on a binary scale. The odour intensity is then expressed in parts per million of butanol. Generally an odour panel may judge odour intensity on a category scale ranging from 0 = no odour to 6 = extremely strong odour.

A measure of odour persistence is often used in conjunction with intensity, since the perceived rate of change in intensity versus concentration is not the same for all odours. This is typically represented as a dose–response curve determined from intensity measurements. The slope of the plot of the log of intensity versus log of dilution ratio is a measure of odour persistence.

The use of a human panel for assessment of odour is highly subjective, costly and time consuming. However, in industries that are associated with foods, beverages, cosmetics, soaps and deodorants, the human panel plays an invaluable role. However, many industrial processes and agricultural operations also produce malodours as well as chemically toxic substances that need to be monitored and controlled. The human panel can only cope with limited numbers of samples, and has to be protected from exposure to potential hazardous substances. Hence, automated methods for odour measurement are desirable.

6.4 Array based sensors for volatile chemicals

Investigations into artificial means of sensing volatile chemicals (an 'electronic nose'), based around the understanding of biological mechanisms of olfaction, started with the identification of suitable materials possessing sensitivity to odorous chemicals. One of the earlier studies in

1950 involved the measurement of surface tension changes in liquids following exposure to an odorant [13].

The first steps to viable odour sensing technology were taken 30 years later, when it was proved that discrimination between odours was possible using a small array of broad specificity semiconductor sensors [14]. The discriminatory power of a small array lies in the utilisation of cross-sensitivities between sensor elements. The responses of the individual sensors, each possessing a slightly different response towards the sample odours, when combined by suitable mathematical methods, can provide enough information to discriminate between sample odours. Since then many research groups, exploiting a variety of sensor technologies, have joined in the development of electronic multi-sensor systems that could eventually start to mimic some aspects of the biological olfactory system. These systems have been given the terminology 'electronic nose' and consist of an array of chemical sensors possessing broad specificity, coupled to electronics and software that allow feature extraction (extraction of salient data for further analysis), together with pattern recognition (identification of sample odour). Software techniques and materials science are important aspects of the development of the system. Advancement in software signal processing techniques, coupled with pattern recognition, enables optimum usage of sensor responses. The specificity and sensitivity of existing chemical sensors are constantly being developed, as well as those of new materials. The heart of any odour-sensing instrument is the type of sensors used for the particular system. The sensors' characteristic responses to the target gases or odours and the operating environment typically restrict applications of the resulting instrument. Many types of sensor have been developed to detect specific gases and vapours since the 1970s. However, it is desired that the sensors in an odour-sensing instrument should have a broad selectivity rather than being specific to one type of gas. The human nose can identify many odours that may contain hundreds of individual chemical components and, therefore, the sensors for an electronic nose should be generalised at the molecular level. The desired properties for sensors are high sensitivity, rapid response, good reproducibility and reversibility to large numbers of chemicals. It is also better for an instrument to be small in size, flexible and able to adapt to many environments, and to operate at ambient temperatures.

Much pioneering work in the use of sensor arrays, before intelligent gas sensing systems were developed, used various multi-sensor systems for identification of different types of gases, often in parallel with analytical methods. Zaromb and Stetter [15] provided a theoretical basis for the selection and effective use of an array of chemical sensors for a particular application. Bott and Jones [16] attempted to build a multi-sensor system to monitor hazardous gases in a mine using six sensors of three different types in combination with oxidising layers and absorbent traps. The system was able to distinguish between gases evolved from a fire and those evolved from diesel engines or explosives. Müller and Lange [17] demonstrated possible identification and concentration measurement of six different gases by means of four MOS gas sensors with layers of different types of zeolite filters. Gall and Müller [18] have adapted similar methods for identifying gas mixtures using partial least squares and transformed least squares methods for analysis, which led Sundgren *et al.* [19] to attempt improvements using three pairs of Pd-gate MOSFETs and Pt-gate MOSFETs. Kaneyasu *et al.* [20] modelled an early version of an electronic nose using six integrated sensor elements with a single chip microcomputer. Odour identifying systems containing two or more different types of sensors are often an attempt to enhance the different dimensional characteristics of the responses. Stetter [21], who was also involved in developing a portable device to detect hazardous gases and vapours, using four different electrochemical sensors, to warn US Coast Guard emergency response personnel [22], demonstrated a combined system with a hydrocarbon sensor and an electrochemical sensor [23, 24]. He managed to get responses from both the hydrocarbon sensor and the electrochemical sensors after passing the vapours through a combustible gas sensor. The resulting data were successfully analysed using pattern recognition methods based on neural networks.

There are numerous types of gas sensor arrays used in electronic noses. The most popular types include metal oxide semiconductor (MOS) sensors, catalytic gas sensors, solid electrolyte gas sensors, conducting polymers, mass-sensitive devices, and fibre optic devices based on Langmuir–Blodgett films. The oxide materials, which have been popular for use in electronic noses, operate on the basis of modulation of conductivity when the odorant molecules react with chemisorbed oxygen species. There are commercially available metal oxide sensors (e.g. from Figaro Inc., Japan) which operate at elevated temperatures, between 100 and 600°C, to help adsorption/desorption kinetics [25]. They are sensitive to combustible materials (0.1–100 ppm), such as alcohols, but are generally poor at detecting sulphur or nitrogen based odours and have a major problem of irreversible contamination with these compounds. Integrated thin-film metal oxide sensors have been designed using planar integrated microelectronic technology, which has advantages of lower power consumption, reduction in size and improved reproducibility; however, they tend to suffer from poor stability. Although there are some oxide materials that show good specificity to certain odours [26], there are a number of advantages in employing organic materials in electronic noses. A wide variety of materials are available for such devices and they operate close to or at room temperature (20–60°C) with a typical sensitivity of around 0.1–100 ppm. Furthermore, functional groups that interact with different classes of odorant molecules can be built into the active material, and the processing of organic materials is easier than that of oxides.

Despite a number of disadvantages including high power consumption, elevated operational temperature, poisoning effects from sulphur-containing compounds and poor long-term stability, MOS gas sensors are the most widely used in gas and odour detection, the main reason being that MOS commercial products have been available for a number of years. Abe *et al.* examined an automated odour sensing system based on plural semiconductor sensors to measure 30 substances [27] and 47 compounds [28]. They analysed the sensor outputs using pattern recognition techniques: Karhunen–Loeve (K–L) projection for visual display output, and the k-nearest neighbourhood (k-NN) method and potential function method for classification. Shurmer *et al.* worked on discrimination of alcohols and tobaccos using tin-oxide sensors based on the correlation coefficient method in their research [29]. Weimar *et al.* demonstrated the possibility of determining single gas components, such as H_2, CH_4 and CO, in air from specific patterns of chemically modified tin-oxide based sensors by using two different multi-component analysis approaches [30]. Most methods applied to identification, classification and prediction of gas sensor outputs were based on conventional pattern recognition techniques until the late 1980s, when artificial neural networks were applied [31]. Gardner and co-workers implemented a three-layer back-propagation network with 12-input and 5-output architecture for the discrimination of several alcohols, where they reported that it was better than the previous work [32] carried out using analysis of variance (ANOVA). Cluster analysis and principal component analysis (PCA) were used to test five alcohols and six beverages from 12 tin-oxide sensors. The results were presented by raw and normalised responses, and showed that the theoretically derived data normalisation substantially improved the classification of chemical vapours and beverages. Further investigations were carried out to discriminate between the blends and roasting levels of coffees, the differences between tobacco blends in cigarettes, and three different types of beer. The result confirmed the potential application in an electronic instrument for on-line quantitative process control in the food industry [33].

Another approach to odour sensing was studied using a quartz-resonator sensor array where the mechanism of odour detection was based on the changes in oscillation frequencies when gas molecules are adsorbed on to sensing membranes. Nakamoto *et al.* employed neural network pattern recognition, including three layer back-propagation and principal component analysis, for the discrimination of several different types of alcoholic drinks using a selection of sensing membranes [34–36].

Persaud and Pelosi [37–40] proposed an odour sensing instrument using conducting polymers after investigating properties of a number of conducting polymers. They have found several organic conducting polymers that respond to gases with a reversible reaction of conductivity, fast recovery and high selectivity towards different compounds. In an experiment with an array of five different conducting polymer sensors and 28 odorants, they observed 20 different sets of responses and showed possible discrimination with 14 of the odorants by measuring changes in the electrical resistance. These results led Persaud *et al.* to produce arrays of 20 gas sensitive polymers that had reversible changes in conductivity and rapid adsorption/desorption kinetics at ambient temperatures when they were exposed to volatile chemicals. The concentration–response profiles of such sensors are almost linear over a wide concentration range to single chemicals. This is advantageous as simple computational methods may be used for information processing. The odour sensing system, developed at UMIST, and commercialised by Aromascan plc, is the successful outcome of this research. The array of sensors has been expanded to 32 sensor elements, and may be expanded further.

6.4.1 Array sensor responses

Figure 6.11 illustrates the dynamic response of a conducting polymer sensor array to acetone vapour, and is representative of the typical behaviour of such sensors. The adsorption/desorption profile is rapid, and each sensor has a different sensitivity to the vapour, some being much greater than others.

Many different kinds of chemicals can be sensed and, as Figure 6.12 illustrates, for ethanol vapour, the amplitude of the response is proportional to concentration. For many single chemicals, an approximately linear response relationship can be modelled. However, for mixtures of chemicals, the response is likely to be non-linear due to competition for binding to adsorption sites.

By normalising the steady state responses of the set of sensors in an array to different volatile chemicals or mixtures of chemicals, unique patterns can be generated, as illustrated in Figure 6.13. This figure shows that it is easy to distinguish visually one pattern from another. The process of discrimination and classification of these multi-dimensional patterns can be automated, and cluster analysis, multivariate analysis and neural networks are commonly used.

The general design of an odour sensing instrument is shown in Figure 6.14. This consists of an array of sensors, the electronics for data acquisition, and a suitable software engine for pattern recognition. The sampling system used to introduce odours to the sensor array is highly important. Many odours consist of complex mixtures and the equilibria between components may change dramatically with temperature. Other factors such as the humidity present in the sample may affect the odour signal recorded by the sensors.

The software provided for a volatile chemical sensing system based on

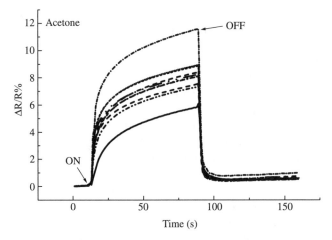

Figure 6.11 *Dynamic response of a subset of conducting polymer sensors to acetone vapour. The y-axis ΔR/R% represents the percent change in dc resistance of the sensor when the vapour is presented to the sensor array*

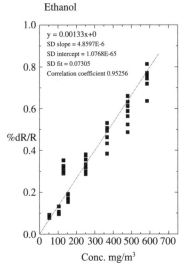

Figure 6.12 *Averaged concentration–response curve from the entire array for ethanol vapour*

conducting polymer sensors consists of three modules: data acquisition and instrument control, data manipulation for extraction of patterns, and pattern recognition.

The acquisition software samples the sensor array resistance at regular intervals, storing the resultant data in the computer. As the resistance of a

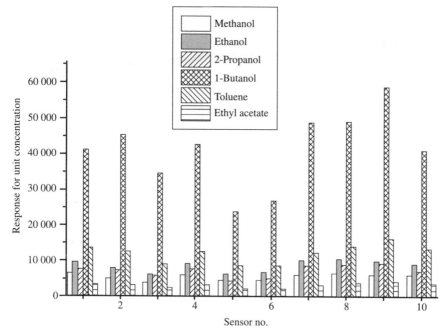

Figure 6.13   *For a selection of ten sensors in a 32-sensor array, the percentage change in resistance per unit concentration ($\% \, mol^{-1}$) is plotted for methanol, ethanol, 2-propanol, 1-butanol, toluene and ethyl acetate. The figure illustrates the wide differences in sensitivity to different substances with different sensors, and also that each sensor responds to some extent to each volatile chemical*

conducting polymer is inversely proportional to temperature, the temperature of the array is controlled and monitored (typically to 35°C ± 0.1°C). The sample temperature and sample humidity are also monitored, since these parameters are important for reproducible sampling. Individual sensors on the array may be deactivated or activated as required. The responses of the sensors are shown in real time in a strip chart display on screen, as seen in Figure 6.11. The signal is expressed as the percentage resistance change of each sensor compared to the initial sensor resistance.

The response of a sensor is then normalised by expressing the fractional change of the individual sensor as a percentage of the fractional changes summed over the whole array, as denoted in eqn. 7.2 for an array of n sensors:

$$N_x = 100 \frac{\Delta r_x / r_x}{\sum_{i=1}^{n} \left| \Delta r_i / r_i \right|} \tag{2}$$

where $1 \leq x \leq n$, N_x is the normalised response of sensor x, Δr_x is the resistance change of sensor x, r_x is the base (initial) resistance of sensor x,

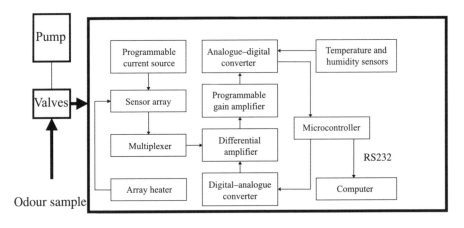

Sampling   Sensing   Data processing

Figure 6.14   A practical design for an array-bassed odour sensing system

and $\Delta r_i/r_i$ is the fractional change in resistance of the ith component of the array. The normalised data thus forms a pattern across the sensor array, as shown in Figure 6.12.

The relevant section of data then can be extracted with the data manipulation software. The data extracted, in the form of a normalised pattern, can be displayed on a screen, exported to spreadsheet programs or exported into the pattern recognition software.

6.5 Practical applications

Using the system described in the preceding sections, it is possible to illustrate some of the applications of artificial odour sensing devices. We have worked on odour sensing devices based on conducting polymer sensor arrays, coupled to suitable sampling systems, hardware and software.

6.5.1 Malodour measurement

Measurement of malodour is very unpleasant for a human panel, and it is desirable to be able to automate such measurements. In order to assess the dynamic changes in odour emission from pig slurry samples, measurements of controlled odour emission from pig slurry were carried out using an odour emissions chamber, and the sensor responses were compared to human panel olfactometric measurements of odour units emitted at discrete sample intervals. In this experiment pigs fed with different diets were being compared. With suitable sensor design it is possible to begin to correlate the sensor response with human panel olfactometric measurements.

In this experiment, an odour chamber was used to collect odour emitted over a period of time from fresh pig slurry. Details of the system used have previously been described by Hobbs *et al.* [41]. Samples of slurry were collected from pigs which had been fed a standard commercial diet (A) and a reduced protein diet (B) for the previous month. Data from pigs that were in the second month of the trial were used. Three replicates of 200 l of each slurry were collected and stored in stainless steel trays ($2 \times 0.5 \times 0.25$ m) with airtight lids in a waterbath at 15°C. An odours and emissions chamber (OEC) produced a controlled environment in which to monitor the release of odorants from slurries. This consisted of an enclosed system of stainless steel ducting (0.5×0.5 m internal section) connected to a variable volume Tedlar air bag. The bag was wrapped around a large plywood roller that ran on inclined rails. Weights added to the roller exerted a pressure on the bag, ensuring that no air entered the system through any leaks. A variable speed fan circulated air at a velocity of 4 m s^{-1} within the temperature controlled OEC. The lower section of the duct was designed to accommodate the tray of slurry, with a variable speed mixing arrangement enabling slurry to be constantly circulated within the tray. The slurry temperature was maintained at 15°C, and air temperature at 20°C to reduce condensation from

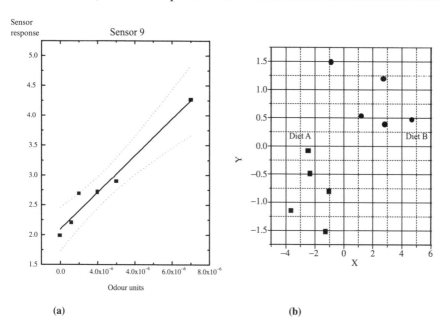

(a) (b)

Figure 6.15 *(a) Correlation of sensor response to pig slurry odour, with odour concentration units measured by a human panel; (b) discrimination between odours from pigs fed different diets A and B using a Sammon Map to show the distribution of odour patterns in a two dimensional space*

the air. Background measurements were made before the slurry tray lid was removed and further samples were taken at 15, 40, 65, 95, 155 and 255 minutes after lid removal.

Figure 6.15a plots the response of the odour sensor system versus the odour units obtained through olfactometry using a human panel. Here for clarity the response of one of the sensors in the array (sensor 9) is shown for pigs fed with diet A. It is clear that a good correlation exists between the sensor response and the human assessment. Using cluster analysis of the patterns obtained, it is seen that effluent from pigs fed with diet A could be differentiated from those of diet B (Figure 6.15b). The results indicate for the first time a good correlation between human perception of odour as measured by olfactometry and the sensor responses. This, coupled with a capacity to discriminate between complex odours, indicates a new future for odour sensing instruments for malodour measurements in many other quality control applications.

6.6 Conclusion

Conducting polymers have a wide variety of applications in gas and volatile chemical sensing. Because the chemistry is readily amenable to attachment of substituent groups on to the polymer backbones, it is possible to create tailored adsorbent materials that are capable of high sensitivity (ppm–ppt) to a large number of chemicals. This chapter has given only a brief description of these possibilities. Much research is focused on the development of sensor arrays that mimic the functioning of the human nose, and with conducting polymers, the inherently high sensitivity to polar compounds allows correlation of sensor response to human response in certain situations. The evolution of this technology will open up a new frontier of objective odour measurements that is a radical change from the traditional chemical analytical methods.

6.7 Acknowledgements

This work was partially supported by BBSRC, EPSRC and Aromascan plc, UK.

6.7 References

1 SKOTHEIM, T. A. (Ed.): 'Handbook of conducting polymers, vols. 1 and 2' (Marcel Dekker, New York, 1986)
2 DIAZ, A. F., KANAZAWA, K. K., and GARDINI, G. P.: 'Electrochemical polymerization of pyrrole', *Journal of the Chemical Society, Chemical Communications*, 1979, pp. 635–636

3 NALWA, H. S. (Ed.): 'Handbook of organic conductive molecules and polymers, vol. 2' (John Wiley and Sons, Chichester, New York, 1997)

4 MARK, H. F., BIKALES, N. M., OVERBERGER, C. G. and MENGES, G.: 'Encyclopedia of polymer science and engineering, vol. 5 (John Wiley and Sons, 1986, 2nd edn.), pp. 462–507

5 KU, C. C., and LIEPINS, R.: 'Electrical properties of polymers' (Hanser, Munich, 1987), pp. 178–285

6 BREDAS, J. L., STREET, G. B., THEMANS, B., and ANDRE, J. M.: 'Organic polymers based on aromatic rings (polyparaphenylene, polypyrrole, polythiophene) – evolution of the electronic-properties as a function of the torsion angle between adjacent rings', *Journal of Chemical Physics*, 1985, **83**, pp. 1323–1329

7 SCOTT, J. C., PFLUGER, P., KROUNBI, M. T., and STREET, G. B.: 'Electron-spin-resonance studies of pyrrole polymers – evidence for bipolarons', *Physical Review B – Condensed Matter*, 1983, **28**, pp. 2140–2145

8 BREDAS, J. L., SCOTT, J. C., YAKUSHI, K., and STREET, G. B.: 'Polarons and bipolarons in polypyrrole – evolution of the band-structure and optical-spectrum upon doping', *Physical Review B – Condensed Matter*, 1984, **30**, 1023–1025

9 MIASIK, J. J., HOOPER, A., MOSELY, P. T., and TOFIELD, B. C.: 'Electronically conducting polymer gas sensors', in ALCACER, L. (Ed).: 'Conducting polymers: Special applications' (Reidel, 1987), pp. 189–198

10 BARTLETT, P. N., ARCHER, P. B. M., and LING-CHUNG, S. K.: 'Conducting polymer gas sensors. 1. Fabrication and characterization', *Sensors and Actuators*, 1989, **19**, pp. 125–140

11 PERSAUD, K. C., and TRAVERS, P. J.: 'Arrays of broad specificity films for sensing volatile chemicals', in KRESS-ROGERS, E. (Ed.), 'Handbook of biosensors and electronic noses' (CRC Press, 1996), pp. 563–592

12 GARDNER, J. W., and BARTLETT, P. N.: 'A brief history of electronic noses', *Sensors and Actuators B – Chemical*, 1994, **18**, pp. 211–220

13 TANYOLAÇ, N. N., and EATON, J. R.: 'Study of odors', *Journal of the American Pharmacentical Association*, 1950, **39**, (10), pp. 565–574

14 PERSAUD, K.C., and DODD, G.: 'Analysis of discrimination mechanisms in the mammalian olfactory system using a model nose', *Nature*, 1982, **299**, pp. 352–355

15 ZAROMB, S., and STETTER, J.: 'Theoretical basis for identification and measurement of air contaminants using an array of sensors having partly overlapping selectivities', *Sensors and Actuators*, 1984, **6**, pp. 225–243

16 BOTT, B., and JONES, T.: 'The use of multisensor systems in monitoring hazardous atmospheres', *Sensors and Actuators*, 1986, **9**, pp. 19–25

17 MÜLLER, R., and LANGE, E.: 'Multidimensional sensor for gas analysis', *Sensors and Actuators*, 1986, **9**, pp. 39–48

18 GALL, M., and MÜLLER, R.: 'Investigation of gas mixtures with different MOS gas sensors with regard to pattern recognition', *Sensors and Actuators*, 1989, **17**, pp. 583–586

19 SUNDGREN, H., LUNDSTRÖM, I., and WINQUIST, F.: 'Evaluation of a multiple gas mixture with a simple MOSFET gas sensor array and pattern recognition', *Sensors and Actuators B*, 1990, **2**, pp. 115–123

20 KANEYASU, M., IKEGAMI, A., ARIMA, H., and IWANAGA, S.:

'Smell identification using a thick-film hybrid gas sensor', *IEEE Transactions on Components, Hybrids and Manufacturing Technology*, 1987, **CHMT-10**, (2), pp. 267–273

21   STETTER, J., JURS, P., and ROSE, S.: 'Detection of hazardous gases and vapors: pattern recognition analysis of data from an electrochemical sensor array', *Analytical Chemistry*, 1986, **58**, pp. 860–866

22   OTAGAWA, T., and STETTER, J.: 'A chemical concentration modulation sensor for selective detection of airborne chemicals', *Sensors and Actuators*, 1987, **11**, pp. 251–264

23   STETTER, J.: 'Sensor array and catalytic filament for chemical analysis of vapours and mixtures', *Sensors and Actuators B*, 1990, **1**, pp. 43–47

24   STETTER, J.: 'Chemical sensor arrays: Practical insights and examples', in GARDNER, J., and BARTLETT, P.: 'Sensor and sensory systems for electronic nose', NATO ASI Series E: Applied Sciences, vol. 212 (Springer-Verlag, Berlin, 1992), pp. 273–301

25   MOSELEY, P., NORRIS, J., and WILLIAMS, D. (Eds.): 'Techniques and mechanisms in gas sensing' (Adam Hilger, Bristol, 1991)

26   EGASHIRA, M., SHIMIZU, Y., and TAKAO, Y.: 'Trimethylamine sensor based on semiconductive metal oxides for detection of fish freshness', *Sensors and Actuators B*, 1990, **1**, pp. 108–112

27   ABE, H., YOSHIMURA, T., KANAYA, S., TAKAHASHI, Y., MIYASHITA, Y., and SASAKI, S.: 'Automated odor-sensing system based on plural semiconductor gas sensors and computerized pattern recognition techniques', *Analytica Chimica Acta*, 1987, **194**, pp. 1–9.

28   ABE, H., KANAYA, S., TAKAHASHI, Y., and SASAKI, S.: 'Extended studies of the automated odor-sensing system based on plural semiconductor gas sensors with computerized pattern recognition techniques', *Analytica Chimica Acta*, 1988, **215**, pp. 151–168

29   SHURMER, H., GARDNER, J., and CHAN, H.: 'The application of discrimination techniques to alcohols and tobaccos using tin-oxide sensors', *Sensors and Actuators*, 1989, **18**, pp. 361–371

30   WEIMAR, U., SCHIERBAUM, K., and GÖPEL, W.: 'Pattern recognition methods for gas mixture analysis: Application to sensor arrays based upon SnO_2', *Sensors and Actuators B*, 1990, **1**, pp. 93–96

31   SLEIGHT, R.: 'Evolutionary strategies and learning for neural networks'. MSc dissertation, UMIST, 1990

32   GARDNER, J., HINES, E., and WILKINSON, M.: 'Application of artificial neural networks to an electronic olfactory system', *Measurement Science Technology*, 1990, **1**, pp. 446–451

33   SHURMER, H., and GARDNER, J.: 'Odour discrimination with an electronic nose', *Sensors and Actuators B*, 1992, **8**, pp. 1–11

34   EMA, K., YOKOYAMA, M., NAKAMOTO, T., and MORIIZUMI, T.: 'Odour-sensing system using a quartz-resonator sensor array and neural-network pattern recognition', *Sensors and Actuators*, 1989, **18**, pp. 291–296

35   NAKAMOTO, T., FUKUNISHI, K., and MORIIZUMI, T.: 'Identification capability of odor sensor using quartz-resonator arrays and neural-network pattern recognition', *Sensors and Actuators B*, 1990, **1**, pp. 473–476

36   NAKAMOTO, T., FUKUDA, A., and MORIIZUMI, T.: 'Improvement of

identification capability in an odor-sensing system', *Sensors and Actuators B*, 1991, **3**, pp. 221–226

37 PELOSI, P., and PERSAUD, K. C.:'Gas sensors: Towards an artificial nose', in DARIO, P., *et al.* (Eds.): 'Sensors and sensory systems for advanced robots', NATO ASI Series F: Computer and System Science (Springer-Verlag, Berlin, 1988), pp. 361–382

38 PERSAUD, K. C., BARTLETT, J., and PELOSI, P.: 'Design strategies for gas odour sensors which mimic the olfactory system', in DARIO, P., *et al.* (Eds.): 'Robots and biological systems', NATO ASI Series (Springer-Verlag, Berlin, 1990)

39 PERSAUD, K. C., and TRAVERS, P.: 'Multielement arrays for sensing volatile chemicals', *Intelligent Instruments and Computers*, 1991, July–August, pp. 147–153

40 PERSAUD, K. C., and PELOSI, P.: 'Sensor arrays using conducting polymers for an artificial nose', in GARDNER, J., and BARTLETT, P. (Eds.): 'Sensors and sensory systems for an electronic nose', NATO ASI Series E: Applied Sciences, vol. 212 (Springer-Verlag, Berlin, 1992), pp. 237–256

41 HOBBS, P. J., MISSELBROOK, T. M., and PAIN, B. F.: 'Assessment of odours from livestock wastes by a photoionisation detector, an electronic nose, olfactometry and gas-chromatography-mass spectrometry', *Journal of Agricultural Engineering Research*, 1997, **60**, pp. 137–144

Chapter 7

Thin film (ClAlPc) phthalocyanine gas sensors

M. E. Azim-Araghi and A. Krier

7.1 Introduction

The technological advances which have revolutionised life during the 20th century have brought with them problems on a world-wide scale. Global warming, environmental contamination and pollution of many types has motivated the search for better detection of gases. Examples of typical requirements might include CO and CO_2 in fire detection, ammonia sensing in agricultural environments, carbon monoxide and nitrogen oxides in car exhausts and sulphur dioxide from industrial pollution. There is also a need for oxygen, hydrogen and humidity sensors for use in a variety of applications.

In elucidating the qualities of a sensor we must first examine the special characteristics we reasonably expect to find in an ideal sensor. Typically these might be sensitivity (often in the ppb range), specificity (ability to recognise a particular gas), selectivity (ability to recognise one gas in the presence of others), rapid response, rapid recovery to 100% of the original state, reproducibility in both response and manufacture, temperature stability, small size, low unit cost and also low power consumption.

Organic semiconductors offer a number of advantages over inorganic materials. Firstly, there is an enormous range and diversity of organic semiconducting materials when compared with the available options amongst inorganic materials. Supplementing this range of basic materials is the ability to modify complex molecules to produce molecular structures with improved electronic materials. A further factor of importance is the inherently small size of the molecular sub-unit, offering the possibility of considerable size reductions in device arrays.

Organic semiconductors may be defined as materials containing a significant number of carbon–carbon bonds which are able to sustain a degree of electronic conduction. They are normally groupable into three classes: molecular crystals, polymers and charge transfer complexes. We will consider in this section only molecular crystals. These are typified by materials such as anthracene, naphthalene and phthalocyanines, which are characterised by weak van der Waals bonding. In these materials intramolecular bonding is stronger than intermolecular bonding and this leads to small molecular overlap, narrow energy bands and low carrier mobility.

Phthalocyanines are extremely stable, both in terms of resistance to chemical attack and thermally. The molecular structure consists of a central metal ion surrounded by benzo-pyrrole units. Most metal phthalocyanines are planar but in some cases (e.g. PbPc) the central ion is displaced from the plane of the molecule. Until the work of Eley [1], the phthalocyanines were regarded as electrical insulators which were useful only as colouring materials or dyes. The observed conductivity is now generally believed to have its origin in the transport of free π ('delocalised') electrons. The electronic interaction between molecules can then be regarded as being due to an overlap of molecular orbitals. The response of delocalised electrons, for example π electrons, in the material is not in itself a necessary prerequisite for high conductivity since, chemically at least, those compounds classified as involving free radicals and charge transfer complexes also exhibit high conductivity.

There are three main mechanisms of conduction which can be used to explain some of the effects experimentally observed in organic semi-conductors. These are based on the band model, the tunnelling model and the hopping model. The dark conductivity, which is of most import-ance to us, seems to be controlled by defects, impurities, surface states, dislocations and grain boundaries. All of these effectively reduce the activation energy for semi-conduction and thereby ensure that the conductivity is extrinsic rather than intrinsic. The factors governing con-duction are usually classified under the general title of traps, which are sub-divided into shallow traps and deep traps. Their distribution in energy is important, and traps distributed uniformly or exponentially with energy have been observed experimentally in various phthalocyanines.

In this section attention will be focused on the factors which influence adsorption of gases on to thin films of phthalocyanines and consequently change their electrical properties. These changes in electrical conductiv-ity can in favourable cases be by factors of up to 10^8. Such huge changes provide a strong basis for the use of phthalocyanines as semiconducting gas sensors.

Conclusions relating to current conduction mechanisms have generally been based on the dc investigation of metal phthalocyanines (current temperature and current voltage characteristics) both in the form of single crystals and thin films [2–4]. In thin films at low voltages the conduction is ohmic while at higher voltage levels space-charge-limited conductivity has been identified as the mechanism responsible for the conduction in samples with ohmic metallic contacts. The ac conductivity in various simple metal phthalocyanines (including NiPc, MoPc, ZnPc, PbPc and ClAlPc) has been shown to vary with frequency f as f^n, where n is an index ≤ 1 [5]. This behaviour normally indicates the predominance of a hopping conduction mechanism at a high excitation frequency [6]. CuPc thin films have been used as the basis of NH_3 and CCl_4 sensors by several workers [7, 8]. A number of phthalocyanines (ZnPc, FePc, CoPc, ClAlPc, etc.) also exhibit a variation in dc conductivity when they are exposed to oxidising gases [9,10].

7.2 Gas sensors

7.2.1 *Ideal sensor qualities*

In elucidating the qualities of a working or practical sensor we must first examine the special characteristics we might reasonably expect to find in an ideal sensor. Typically these might be:

- sensitivity
- selectivity
- short response times
- reproducibility
- simplicity/low cost
- ruggedness and portability.

7.2.2 *The FET sensor*

There has been considerable interest and emphasis placed on the use of so called transconductance mode sensors. These are sensors which are based on the well-known FET structure and include MISFETs [11], ISFETs [12] and various CHEMFETs [13], all with commercial viability.

The semiconductor surface potential in devices based on the FET principle is modified by potential variations elsewhere in the structure, usually at an insulator–electrolyte interface. Emphasis has been directed towards measuring the concentration of ions in solution; for example, H^+, (pH), K^+, Na^+ and Ca^{2+} in blood. These devices are categorised as ISFETs. A pd-gate MISFET hydrogen sensor and a pd-MOS hydrogen sensor [14] have been reported, as well as sensors using other metals.

7.2.3 *Metal oxide sensors*

Sensors based on the change in conductivity of an inorganic semi-conducting oxide on exposure to certain gases have been studied extensively [15]. Generally these are based on the compound SnO_2 [16,

17], and the electrical conductivity arises from non-stoichiometric oxide formation. Metal oxide gas sensors have many of the qualities required of an ideal sensor. For example, they are rugged, do not require complex electronic backup circuitry, have low level toxic capabilities and are very sensitive. However, they do not yet exhibit long term stability, reproducibility, short response time and, more importantly, selectivity.

7.2.4 Langmuir–Blodgett films

In the last decade there has been great interest shown in the use of Langmuir–Blodgett films for device applications. In particular, the potential for using Langmuir–Blodgett (LB) phthalocyanine films as gas detectors has been noted [18]. The benefits of using LB films over amorphous or polycrystalline films include accurate thickness control, increased stability and high breakdown strength. One must bear in mind that fabrication of LB films is far from straightforward compared with vacuum sublimation and is most likely to meet specialist applications where small runs are economically feasible.

7.2.5 Thin film gas sensors

The need for small, economic solid state thin film gas detectors is now widely recognised. Their main application is likely to be in replacing bulky, expensive gas detection systems which are available at present, with the added criterion of being suitable for use in a variety of applications for industry and monitoring of pollutant gases.

7.2.6 The rationale of gas sensor characterisation

When considering the thin film metal phthalocyanine type of sensor, there are a number of behavioural trends which are already well-established, i.e. thermal stability, sensitivity, reversibility, low cost, etc. Maximum sensitivity occurs at different temperatures for different gases and different metal phthalocyanines. The objectives of any gas sensor design are to establish the optimum operating conditions and the reproducibility and reliability of the sensor. This involves determining the spread in the various operating parameters for different metal phthalocyanine sensors under different operating conditions and also finding out the stability of these parameters. The characteristics of any phthalocyanine based gas sensor are likely to be highly temperature dependent and thus all measurements should be carried out at a number of temperatures. The temperature range of interest will have to be different for different metal phthalocyanines and for different gases. There is evidence [19] that the conductance of some devices, especially at low temperatures, requires fairly long times to reach a steady state. Hence, measurements must be made only after a sufficient time has been allowed at a given temperature for the conductance to reach a steady value.

In the testing procedure, the importance of the integrity and reproducibility of the gas mixtures cannot be overemphasised. Extreme care must be taken in designing and constructing the gas flow and mixing systems. The rate of access of the gas mixture to the sensor and the integrity of that gas mixture can be very dependent on the design of and the material used in constructing the enclosure. It is usually advantageous to minimise the volume of the enclosure, to allow the gas to flow through the enclosure past the sensor, and to ensure that the material of which it is made does not either decompose at the temperature to which it is raised by the sensor or strongly adsorb some gases, since both adsorption of analyte gas and desorption of contaminants can lead to erroneous data.

7.2.7 The measured parameters

Conductance: The principal parameter measured is the resistance of the phthalocyanine, whether it is in thin film, single crystal or any other physical form. Since it is unlikely that the conductivity will be uniform throughout the material, and because the measured parameter is in fact a surface resistance, it is meaningless to compute the absolute parameters of conductivity or resistivity. There can be instances, however (for example, in the study of single crystals), where the use of an average conductivity can eliminate some effects attributable to different crystal sizes. Essentially all the operating characteristics of the sensor are derived from this one measurement. This is perhaps both the strength and the weakness of phthalocyanine sensors: a strength because it is a simple, easily measured parameter, but a weakness because it is a second-order parameter which, although very sensitive to some reactions at the solid surface, is not a good indicator of the exact processes taking place.

Sensitivity: More explicitly, sensitivity (s) can be defined as (conductance in gas − conductance in air)/(conductance in air) expressed as a percentage, i.e. $(\Delta\sigma/\sigma_{air}) \times 100\%$, which is the easiest parameter to handle when considering adsorption of donor gases on n-type oxides, since in these cases $(\Delta R/R) \times 100\%$ (where R is the resistance) tends asymptotically to 100%. This definition of sensitivity is fairly arbitrary and other parameters, such as the change in conductance for a given concentration of gas ($\Delta\sigma$), can in some instances be, arguably, a better measure. In some circumstances, the value of ($\Delta\sigma$/concentration) can also have some advantages, although the $\Delta\sigma$ variation with concentration is only linear over small concentration ranges. The sensitivity to the gas of interest must be measured over the whole of the temperature range. This gives most of the parameters required to characterise the sensor fully. Two such parameters which can be used as bench-marks for sensitivity comparison between sensors or between gases are the maximum sensitivity and the temperature at which it occurs. From the data in different gases, an indication of the selectivity with respect to one gas relative to any of the other gases at any point in the temperature range can be determined. The selectivity is usually defined as the (sensitivity to gas 1)/(sensitivity to gas 2) for equivalent concentrations of both gases, or, sometimes, for the concentration of the gases known to be involved in the application of interest. Thus, from the study of sensitivity, an optimum temperature or range of temperatures over which the selectivity is acceptable can be determined.

Response time: In gas detection this is usually defined as the time taken to achieve 90% of the final change in conductance following a step change in gas concentration at the sensor. The response time is an important parameter since its value can determine the applicability of the sensor; unfortunately it is probably the most difficult of all the operating parameters to measure. It will frequently require special gas flow systems, customised for particular gases to ensure that the step change in gas concentration (this usually from clean air to a given concentration) is faster than the response time of the sensor. This can present problems, especially when dealing with highly adsorptive or reactive gases.

Stabilisation time: This is the measure of the time taken for the conductance to reach a steady state, in dry air (20% O_2) or a gas mixture, after the sensor is switched to operating temperature from cold. This parameter is important if the device is to be used in an intermittent mode, i.e. it is only raised to the operating temperature when a measurement is required. It is much less important when the device can be maintained permanently at the operating temperature, in which case it only governs the length of time required, following switch-on, before a measurement can be made. This parameter can often limit the use of phthalocyanine devices since, at low temperatures, the rates of desorption of some adsorbed species can be slow and the time taken to achieve a steady state therefore precludes intermittent operation. However, at high temperatures such an operation is sometimes feasible.

Some of the information obtained in a full characterisation could be relevant to elucidating the mechanisms involved in gas adsorption at the phthalocyanine surface and the consequent conductance changes. Parameters such as activation energies of conduction in the various temperature ranges can be derived. Such information must, if only indirectly, provide some insight into the processes occurring at the surface.

7.3 Experimental details

In order to investigate the electrical (dc and ac), optical and gas sensing properties of phthalocyanines (Pcs), typical interdigital (planar) and sandwich devices may be adopted. The experimental and analytical methods employed in this regard are categorised as follows:

- synthesis and purification of the material under investigation, i.e. chloroaluminium phthalocyanines (ClAlPc)
- structural studies – differential scanning calorimetry (DSC), thermogravimetric analysis (TGA), scanning electron microscopy (SEM) and X-ray diffraction [20]
- fabrication of interdigital and sandwich (Au–Pc–Au) devices by means of photolithography and thermal evaporation techniques
- electrical measurements in vacuum for determining different electrical parameters usable in gas sensing analysis.

7.3.1 Device fabrication

Various techniques can be used for the preparation of thin films of semiconductors, metals and insulators. These include vacuum evaporation, sputtering, chemical methods and ion-beam techniques, etc. To prepare a film of controlled composition and homogeneous structure it needs to be deposited at a controlled rate under an accurately monitored ambient gas pressure. Details of different techniques of deposition and uses of the films so prepared can be found elsewhere [21–23].

The technique used in this work for the preparation of planar devices is vacuum deposition by thermal evaporation. The thermal evaporation technique consists of vaporisation of the solid material by heating it to a sufficiently high temperature and condensing it at 10^{-5} mbar or less. The heating of the material can be carried out directly or indirectly (via a support) by a variety of methods. The most common method is to support the material by a filament, basket, spiral or boat, which is heated electrically. The choice of the filament or boat material is primarily determined by the evaporation temperature and the resistance to alloying and/or chemical reaction with the evaporated material. The advantages of thermal evaporation in vacuum are:

- the material vaporises at lower temperatures
- the deposition of the film is more uniform in thickness.

Planar gas sensor devices: Au–ClAlPc–Au planar devices with interdigital electrodes were prepared in order to investigate the dark conductivity and the gas sensitivity of (ClAlPc) chloroaluminium phthalocyanine. The interdigital gold electrodes were in an eight-finger configuration with a 100 μm gap so as to maximise the exposed Pc surface and to allow the achievement of high electric fields (E) from low voltages (to minimise the power consumption). The gold electrodes were similarly produced by thermal evaporation. The fabrication of a complete sensing device includes:

- substrate preparation
- electrode fabrication
- phthalocyanine deposition.

The nature and surface cleaning of the substrate are extremely important because they greatly influence the properties of the films deposited on to them. Glass, quartz and ceramic substrates are commonly used for film deposition.

For the purpose of this work, polyborosilicate glass slides were used as substrates for gas sensor devices, mainly because of the fact that polyborosilicate is electrically insulating, shows a reasonable thermal conductivity and is available in the form of slides. To obtain coatings that adhere strongly to the substrate and avoid the peeling of the film from it, the substrate must be free from contamination such as grease, dust, etc. Polyborosilicate glass slides, free from scratches, of dimension 25×25 mm were subjected to the following cleaning procedure:

1. The substrates were mechanically polished with precipitated calcium carbonate in a $1:5$ solution of Decon-90 in de-ionised water.
2. The slices were rinsed by spraying with de-ionised water.
3. Subsequently, a set of five slides were mounted in a PTFE holder and cleaned ultrasonically in de-ionised water for 40 min, then blow-dried.
4. Finally, they were removed and stored in a vacuum desiccator (Figure 7.1) until required.

Interdigital electrode preparation: The most common method of preparing thin metal films is by thermal evaporation in a vacuum. By reducing the pressure of the gas atmosphere, the chemical reactions between the material to be evaporated and the residual gases are reduced, and as the gas pressure is lowered the average distance travelled by vapour molecules before colliding with a gas molecule increases (i.e. the mean free path of the molecule in the residual gas increases). To get the best specimen for thin film investigations the ultimate vacuum should be as low as possible. However, a certain amount of residual gas may still exist, such as N_2, CO, CO_2, H_2O and some hydrocarbon products (from rotary and diffusion pumps).

A conventional oil diffusion pumped vacuum system was used for the thermal evaporation of the interdigital gold electrodes. A liquid nitrogen trap was used to reduce the amount of hydrocarbon vapours present in the residual gas. After approximately 20 h pumping, this system was capable of creating a pressure of 10^{-6} mbar in the evaporating chamber.

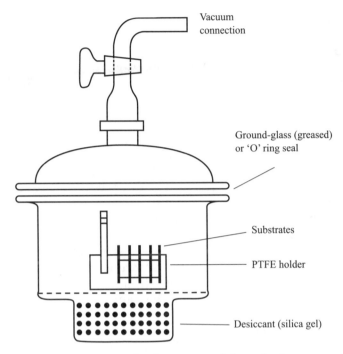

Vacuum
connection

Ground-glass (greased)
or 'O' ring seal

Substrates

PTFE holder

Desiccant (silica gel)

*Figure 7.1   Vacuum desiccator for the storage of cleaned substrates and
            fabricated devices*

To monitor the evaporation, feed-throughs were connected to the evaporation chamber, filament heater, film thickness monitor, shutter manipulation and electrical measurement leads. Gold was used for the device electrodes (800–1200 Å), mainly because it is known to achieve the best ohmic contact with phthalocyanines [24]. The gold was evaporated at a pressure of 10^{-6} mbar from a molybdenum boat. Prior to gold evaporation, the molybdenum boat was degassed over a period of 90 min. A quartz film thickness monitor was installed to control the film deposition rate and to indicate the final thickness. The substrate and the thickness monitor probe were placed approximately 12 cm above the source. When the vacuum pressure was stabilised the evaporation was carried out: the current through the filament was slowly increased while maintaining an evaporation rate of about 1 Å s^{-1}. To prevent contamination by volatile impurities and water vapour and to keep spattering to a minimum, a movable shutter was kept over the source (boat) until the first 100 Å of material had been sublimed. The source was then uncovered and deposition took place. Finally, the coated substrates were left for 2 h to cool down, and then they were removed from the vacuum chamber for photolithography.

Photolithography: The coated substrates were used to fabricate inter-digital electrodes with a spacing of 100 μm as shown in Figure 7.2 below. Because of the small size of the inter-electrode spacing, photolithography was used instead of a simple evaporation mask. The photolithography procedure consisted of the following steps, as indicated in Figure 7.3:

1. The substrates were baked in air at 200 °C for 30 min to desorb volatile contaminants, especially water vapour. After cooling down the substrates, they were then coated with photoresist AZ 1350 (about 1 cm³ of resist was deposited on to the substrate and spun for 20 s at 350 rpm to ensure an equal spread of resist). Then they were soft baked at 80°C under an infrared heater for 25 min.
2. After cooling, they were exposed to ultraviolet light through a positive contact mask for 18 s.
3. The exposed part of the photoresist was removed by developing using A–Z developer for 1 min. Subsequently, the samples were rinsed in de-ionised water until all unwanted resist had been removed, and then blow-dried. The samples were post-baked at 120°C for 30 min.
4. The unwanted material was then removed using a special gold etchant (gold etchant consists of 41% hydrochloric acid (HCl), 9% nitric acid (HNO_3) and 50% de-ionised water (H_2O)).

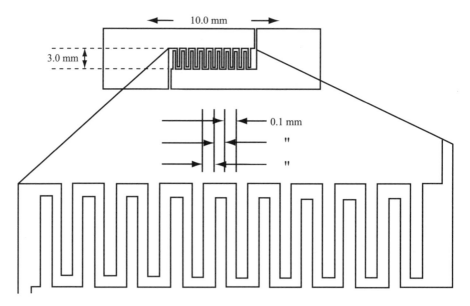

Figure 7.2   Diagram showing geometry of interdigital electrodes

5. Finally, the electrodes were washed with acetone, then rinsed in de-ionised water to remove the remaining unexposed photoresist.

At this stage the electrodes were split into individual devices ready for phthalocyanine deposition (Figure 7.2). Finally, a highly conducting silver paint (Electrodag 915) was used to connect fine copper wires for external contacts. The electrodes were then stored in a vacuum desiccator.

Sandwich devices: Au/ClAlPc/Au can also be prepared to enable conductivity measurements to be carried out to determine conduction mechanisms. The substrate used for making the sandwich devices was an ordinary 5×5 mm^2 polyborosilicate glass slide. Prior to use, the glass slides were thoroughly cleaned to remove contamination, which mainly consists of dust particles or a layer of hydrocarbons from the atmosphere. Gold electrodes of thickness 800–1200 Å were evaporated to form both bottom and top ohmic contacts to the phthalocyanine film. A sequential masking system was used so that sandwich structures were formed without breaking the vacuum.

Phthalocyanine deposition: The gold electrodes were coated with purified Pcs to produce amperometric gas sensing devices using the vacuum sublimation technique. Evaporation was normally carried out with the

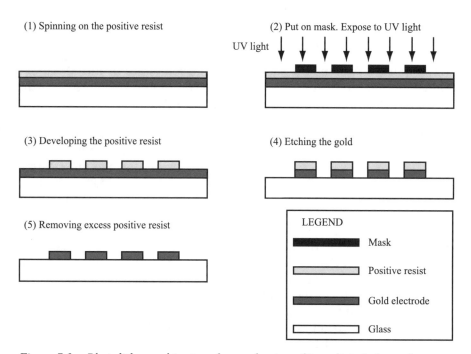

Figure 7.3   Photolithographic steps for production of interdigital electrodes

substrate at a distance of 10–15 cm from a molybdenum boat. A fine gauze was placed over the molybdenum boat to overcome the spattering of ClAlPc material during evaporation, which is related to gas desorption from the material when it sublimes. The sublimation was carried out in a vacuum of about ~10^{-6} mbar. ClAlPc thin films were deposited on to patterned inter digital substrates in the temperature range of 300–570 K. The deposition rate was 0.1–0.2 nm s^{-1}. The evaporation rate and the substrate temperature are of crucial importance in the determination of the subsequent film structure [25–27]. To produce the Pc films in the α-phase (or the monoclinic form in the case of ClAlPc) the substrates have to be maintained at room temperature during deposition, whereas for β-phase (the triclinic form for ClAlPc) the substrates are to be kept at about 550 K. The substrates were heated using a simple heater consisting of a nichrome wire wound round a mica sheet and then sandwiched between two copper plates.

The current through the molybdenum boat was increased slowly in order not to disturb the vacuum pressure. A shutter was used to cover the substrates at the beginning of the sublimation to prevent contamination due to impurities with lower sublimation temperatures. When the required thickness was reached the substrates were covered to prevent further deposition. The devices were left for about 2 h to cool down. They were then removed from the vacuum chamber and stored in a vacuum desiccator.

7.3.2 Measurement techniques

Test chamber: A standard vacuum system consisting of an oil vapour diffusion pump backed by a rotary pump was modified and used for the measurements. The conductivity of the phthalocyanines is known to be affected by the ambient gases [28]; therefore, experiments were carried out in a vacuum of 10^{-6} mbar throughout the work. The chamber pressure was monitored using an Edwards DO 43-11 Penning gauge. The steel base of the vacuum chamber had the facility for making external electrical connections (i.e. for heater, devices, thermocouple, etc.). A needle valve was also provided at the base for bleeding the required gas into the chamber.

The potential application of Pc thin films as toxic gas sensors is based on the change of the electrical properties, especially dc conductivity. It is known that the electrical properties of halogenated phthalocyanine films are affected by the presence of an active gas as well as many other factors such as light, which induces photogenerated electron–hole pairs, mechanical vibrations creating energetic phonons causing further carrier scattering, pick-up, etc. Therefore, in order to ensure that the monitored gas–Pc reaction is solely due to the presence of the active gas in ambient, precautions have to be taken to reduce to a minimum the influence of other factors. In order to eliminate both the possibility of photoelectric effects and interference with other types of pick-up, the glass vessel of the test chamber was either covered with aluminium foil for shielding from all incident light, or alternatively an earthed stainless steel chamber was used. Short co-axial screened cables equipped with BNC 50 Ω connectors were used for all connections and were fastened to the bench to avoid unnecessary movement. Measuring cables were physically isolated from power cables supplying mains or heating current. The voltage applied across the devices was obtained from a dc power supply unit (Weir 4000 series). Thermofoil heaters (Minco HK-913) were used to adjust the sample temperature which was measured using an alumel/chromel thermocouple placed in the vicinity of the sample.

Current–voltage measurements: The choice of measuring instrumentation depended upon the resistivity of samples under test and the accuracy required. For low resistivity samples, the bridge method or simple voltmeter–ammeter type experiments were suitable, whereas for the high resistivity samples used here, high input impedance electrometers were more appropriate (Keithley type 610 B and 610 C). In order to measure the current accurately, the output from the rear panel of the electrometer was fed to a digital multimeter (Keithley 177 microvolt). Dark current–voltage characteristics were measured under two different configurations. In the first configuration the samples were mounted in the vacuum chamber described earlier. The devices were heated by placing them on the thermofoil heater supplied by a stabilised power supply. The temperature of the devices was measured using a platinum/iridium thermocouple placed in the vicinity of the sample. In order to cool the substrate, a hollow copper block vented by two copper pipes (diameter 1 cm) was fitted in the centre of the vacuum chamber. The other ends to the two pipes were vented out through the vacuum seals and were used as an inlet and an outlet for the circulation of liquid nitrogen. Thus low temperatures in the range of (–80°C to 0°C) were maintained during some of the experiments. All measurements were taken after the devices were allowed to stabilise in a vacuum at 10^{-6} mbar or better. In the second configuration, the devices were simply mounted in a light-tight PTFE chamber shielded with a stainless steel box as shown in Figure 7.4.

Temperature measurement: For thermal measurements, a thermofoil heater (Minco HK-913) which was powered by a dc power supply, was placed directly beneath the devices block inside the test chamber. Good thermal contact between the devices was ensured by using a high vacuum grease (Dow Corning). Temperatures up to 480 K were achieved by gradually increasing the voltage across the heater. The following

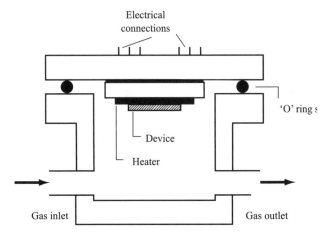

Figure 7.4   The PTFE chamber experimental testing set up

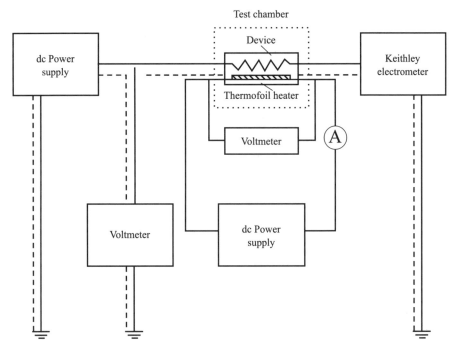

Figure 7.5   Block diagram of the electrical measurement circuit for I-V-T characteristics

temperature measurements were made using the circuit arrangement as shown in Figure 7.5:

- *current–temperature measurements (I–T)* at fixed voltage
- *current–voltage measurements (I–V)* at several fixed temperatures.

7.3.3 Gas sensor testing

Interdigital planar electrodes with the phthalocyanine film as the active layer were chosen to study the sensitivity of phthalocyanines and their response to toxic gases. The effect of various gases was examined under different conditions of temperature, pressure, gas concentration, etc. The devices were brought into direct contact with the gas to be tested and the reaction was then recorded. The adsorption of gases when carrier transfer is involved (chemisorption) generally leads to a measurable conductivity variation. Oxygen is known to be adsorbed strongly at (ClAlPc) phthalocyanine surfaces and to diffuse into ClAlPc films under certain circumstances. Zero grade air (20% O_2) was usually used as a reference gas for device stabilisation and as a carrier gas (to dilute gases such as NO_2, NH_3 and Cl_2), so as to simulate the test conditions likely to be met by an operating gas sensor. In order to investigate the mechanism of gas adsorption on to Pc surfaces, we should consider two situations:

- adsorption on to a completely clean surface (i.e. device in vacuum)
- adsorption on to pre-covered surface (air ambient or zero grade air).

Thus two different types of gas exposure were adopted for electrical measurements (dark current).

Gas flow system (dynamic): The set-up shown in Figure 7.4 was used. Measurements were carried out at atmospheric pressure at room temperature and also at several different temperatures up to 480 K. The gas cylinders were connected to the gas handling system and test chamber using needle valves and flow gauges. All connecting pipes were made from 316 stainless steel with stainless steel fittings. (Alternative connections of PTFE could be used.) All the gases used during these experiments were supplied by BOC Special Gases and were used without further purification.

Static gas sensor testing system: The static gas sensor system described in 7.3.2 was used for the experiments performed 'in situ' in vacuum. The patterned interdigital electrodes were mounted in the vacuum chamber (10^{-6} mbar) and then the device was prepared by sublimation of the phthalocyanine (ClAlPc). Then, without breaking the vacuum the electrical circuit was connected to the system and the device was left to stabilise during the night under a fixed applied bias, and the gas inlet valve was opened, allowing the gas to fill the test chamber. The partial pressure of the gas was monitored using a Pirani gauge. Normally, zero grade air was introduced to the test chamber at atmospheric pressure. The variation of the dark current of the device at a fixed bias was then measured until it stabilised. After taking appropriate measurements, the inlet valve was closed and the chamber was then re-evacuated.

7.4 Experimental results

7.4.1 Variation of the dark current with temperature

Typical results of the variation of device current (log I) with inverse temperature ($1/T$) at a fixed voltage under forward bias are shown in Figure 7.6 for a ClAlPc thin film device in high vacuum in the dark. An applied voltage of 0.6 V was chosen so that the devices were always operating in the ohmic region of the I–V characteristic. The I–T measurements were taken during both the heating and the cooling of the thin film device. The data of Figure 7.6 were taken in a vacuum immediately after film deposition, so that conduction arising from impurity levels due to in-diffusion/adsorption of oxygen and water vapour on to the ClAlPc was minimised. This was confirmed by the high degree of linearity observed in the plots and the reproducibility observed between heating and cooling cycles. In the present case, the situation is considered quantitatively on the basis of existing theory by Meier [29], assuming that the dominant levels are conduction and valence bands. The generalised expression of the conductivity for the semiconductor materials in terms of the temperature and band-gap $\Delta E = E_c - E_v$ (i.e. E_c and E_v are the conduction and valence band-edge, respectively) is

$$\sigma = \sigma_0 \exp(- \Delta E/2k_B T) \tag{1}$$

The slope of Figure 7.6 gives the width of the band-gap and thereby the activation energy $E = \Delta E/2$. The activation energy parameter determined from the data in Figure 7.6 is 0.65 eV.

One of the ClAlPc thin film gas sensors was stabilised in zero grade air (<3 ppm (vol) H_2O, <1 ppm CO_2) at room temperature for several hours. Then the temperature of the device was raised from room temperature to 500 K in steps of 10°C, each temperature step being 50 ± 4 s long. The conductivity of the ClAlPc thin film was recorded at the end of each temperature step throughout the heating procedure. The relation between the dark current and the temperature is shown in

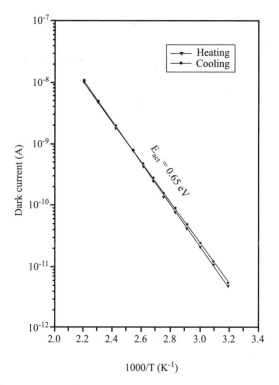

Figure 7.6 The variation of current with temperature for a ClAlPc thin film heated and cooled in high vacuum under dark conditions and in the ohmic region

Figures 7.7 and 7.8 for the ClAlPc thin film sensor on exposure to NO_2 and O_2. We observed that under the present experimental conditions the dark current goes through a maximum at a temperature $T_{max} = 90$ °C.

At $T < T_{max}$ the current increases with temperature. If we assume that ClAlPc is a sufficiently good semiconductor so that we can apply band theory, then we can consider the ClAlPc film to be effectively doped by the adsorbed gas (NO_2 or O_2) which forms an acceptor level within the ClAlPc band-gap. The film conductivity is thermally activated according to the equation, $I = I_0 \exp(-E_{act}/k_B T)$, where $E_{act} = 0.15 \pm 0.01$ eV and 0.18 ± 0.01 eV for ClAlPc doped with NO_2 (500 ppm) and O_2 (20%), respectively, according to our measurements in Figure 7.8.

At temperature $T > T_{max}$ the current decreases with increasing temperature. The higher temperature causes desorption of the adsorbed molecules and consequently a decrease in the number of extrinsic charge carriers and hence in the conductivity of the ClAlPc thin film sensor.

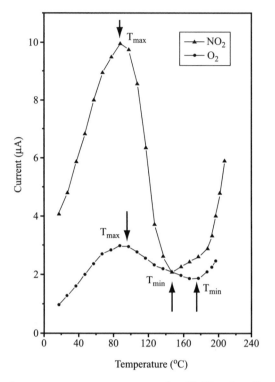

Figure 7.7   Dark current versus temperature for ClAlPc thin film gas sensor

At temperatures > 470 K the dark current increases more rapidly with temperature. This may be considered as non-extrinsic behaviour and is due to thermal activation energy of the non-extrinsic charge carriers. E_{act} is found equal to 0.55 ± 0.01 eV and 0.27 ± 0.01 eV for ClAlPc doped with 500 ppm NO_2 (in the presence of 20% O_2) and 20% O_2, respectively. In fact both thermal activation (non-extrinsic conductivity) and gas desorption are active phenomena over the whole temperature range of 290–500 K, but at temperatures $T < T_{max}$ the semiconductor effect is dominant, whereas at temperatures $T > T_{max}$ desorption is dominant and this decreases the conductivity. Similar results have been reported for NO_2 in copper phthalocyanine and for chlorine in PbPc [30].

To explain further the information about localised levels created in the ClAlPc band-gap as a result of adsorption of NO_2, the I–T data of Figure 7.8 were analysed within the framework of the Roberts and Schmidlin model [29]. By using this model the extrinsic conductivity of the ClAlPc can be expressed in terms of the band-gap and the absolute temperature as

$$\sigma_{ex} \, \alpha \, \exp(E_v - E_h)/k_B T \tag{2}$$

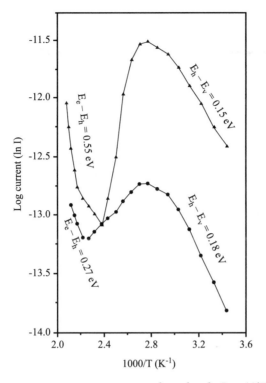

Figure 7.8   The current–temperature curve plotted as ln I vs (1/T)

where E_h and E_v are the dominant hole (acceptor) level and valence band-edge energies, respectively. Hence the activation energy for extrinsic conduction is thus $E_{ex} = E_h - E_v$. The corresponding non-extrinsic (pseudo-intrinsic) conductivity σ_{pi} and activation energy E_{pi} are

$$\sigma_{pi} \, a \, \exp[(E_v - E_h + 1/2(E_h - E_e)]/k_B T \tag{3}$$

and

$$E_{pi} = (E_h - E_v) + (E_e - E_h)/2 \tag{4}$$

respectively, where E_e is the dominant electron (donor) level. For p-type materials if the excess acceptor concentration is small in comparison with the concentration of holes excited from the dominant electron level to the dominant hole level, then the Fermi energy is located between the two dominant levels in such a way as to make the concentration of holes in the dominant hole level equal to the concentration of the electron dominant electron level. This compensated behaviour is termed non-extrinsic or pseudo-intrinsic [31].

The linearity of Figure 7.8 at $T < T_{max}$ and $T > 195°C$ may be attributed to extrinsic and non-extrinsic (pseudo-intrinsic) conductivity, respectively. The region $T < T_{max}$ gives a satisfactory fit to eqn. 7.2 and the activation energy of 0.15 eV relates to the ionisation energy of the NO_2 induced acceptor levels. This is slightly less than the value of 0.18 eV obtained for O_2. The region $T > 195°C$ may be described by eqn. 7.3. Substituting the experimental data obtained from regions $T < T_{max}$ and $T > 195°C$ in Figure 7.8 into eqns. 7.2–7.4, $(E_h - E_v)$ and $(E_e - E_h)$ are found to be 0.15 eV and 0.8 eV, respectively. The temperature T_{max} of the conductivity peak was determined for ClAlPc doped with NO_2 and O_2 (Figure 7.7). T_{max} represents a relative measure of the bonding energy between the gaseous doping molecules and the adsorption sites of the phthalocyanine film, (although this energy cannot be determined exactly from the value of T_{max}), which depends on the experimental conditions. The temperature T_{max} which gave the maximum conductivity of the phthalocyanine film was determined by subjecting the doped film to increasing temperature steps. T_{max} could be used as a design parameter for a thin film gas sensor, and the temperature at which it occurs could provide information on the concentration and the nature of the gas that the phthalocyanine film is exposed to. Furthermore, we expect the maximum response at this temperature.

In the present work we also examined the reversibility of ClAlPc to NO_2 exposure at several different temperatures (300 K, 350 K, 380 K and 425 K). After the test device had reached a stable state at a required temperature with a flow of zero grade air (which normally took about 2 h) then a 15 min pulse of NO_2 was introduced to the test chamber. Figures 7.9a and 7.9b show the response and recovery of these devices. As can be clearly seen, the response increases with increasing temperature up to 350 K. Then we observed that, by increasing the temperature, the sensitivity was reduced, which is in general agreement with findings by other workers [28]. In Figure 7.7 we also observed a minimum (T_{min}) in dark conductivity near 140°C for the sensor in NO_2 (500 ppm) and near 180°C for the same sensor in O_2 (20%). Although operation of the sensor at this minimum results in slightly lower sensitivity, Figure 7.9b shows that the sensor has a very good reversibility at this temperature. At high temperatures the adsorption and desorption processes accelerate and the response of the film becomes more reversible (Figure 7.9b). As is well known, the dark current actually increases from a very low value ($10^{-10} - 10^{-11}$ A) to about $10^{-4} - 10^{-5}$ A at room temperature on first exposure of the film to the test gas. Subsequent exposure (using dry air as the ambient) does not result in a complete release of the gas even if the ClAlPc film is heated to 220°C and it is necessary to cycle the films thermally in a vacuum if a complete recovery is required. Whereas T_{max} is the optimum temperature for sensitive gas detection, our results show that a

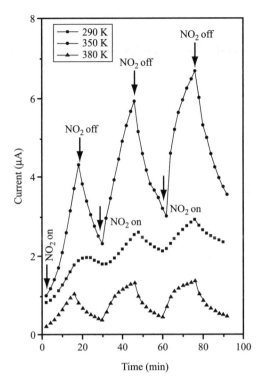

Figure 7.9(a) *The variation of current vs time for ClAlPc thin films exposed to three cycles of 500 ppm NO_2 at 290 K, 350 K and 380 K*

good reversibility is not obtained unless the ClAlPc sensor is operated at or above the minimum temperature, T_{min}.

7.4.2 *Sensitivity*

Our experimental data on ClAlPc gives some important information on the use of ClAIPc as a gas sensor. From Figures 7.7 and 7.8 it is clear that 20% O_2 induces a smaller conductivity increase than 500 ppm NO_2 and that in fact 500 ppm NO_2 could be easily detected in the presence of 20% O_2 (zero grade air). However, O_2 desorbs at a slightly higher temperature than NO_2 (because O_2 molecules are more strongly bound to the ClAIPc surface, as shown in Figure 7.10), reducing the sensitivity of the ClAIPc film to NO_2 compared with a film having a clean surface. The curve of Figure 7.9b leads us to speculate that the response of ClAIPc films to NO_2 has two parts: (i) the fast initial change in current which may be due to NO_2 adsorption on easily accessible adsorption sites from which oxygen is easily displaced; and (ii) the slower change in

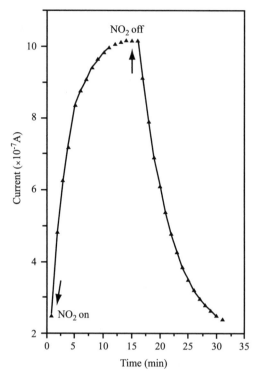

Figure 7.9(b)   The variation of current vs time for ClAlPc thin films exposed to 500 ppm NO₂ at 425 K

conductivity which is caused by the replacement of O_2 by NO_2 molecules on strong adsorption sites. This is again in general agreement with the findings of previous workers using other phthalocyanines [32]; ClAlPc exhibits a high sensitivity to NO_2. More explicitly, sensitivity (s) can be defined as

$$s = (\Delta I/I_0)/(\Delta C/C_0) \qquad (5)$$

where ΔI is the increase in dark current I_0, produced by a change in adsorbed gas concentration (ΔC) from an initial value C_0. The sensitivity of the ClAlPc sensors was measured for several different concentrations of NO_2 in zero grade air at three different temperatures around the maximum temperature T_{max}. Graphs of conductivity versus time for ClAlPc films exposed to 100, 200 and 300 ppm NO_2 at three different temperatures are shown in Figures 7.11a, 7.11b and 7.11c. In each case the sensitivity of the sensor was calculated from the conductivity curves. Clearly the sensitivity of ClAlPc to NO_2 decreases when the temperature increases; this result is in good agreement with findings of previous workers [10]. Values of the sensitivity of ClAlPc films to NO_2 at tempera-

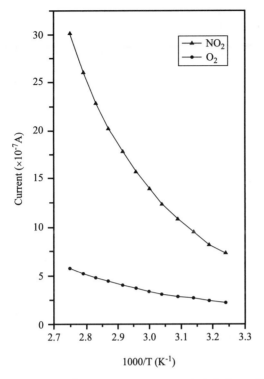

Figure 7.10 *The variation of current vs temperature (V_1) for a ClAlPc thin film after exposure to NO_2 (500 ppm in zero grade air) and O_2 (20% in zero grade air)*

tures 390, 420 and 450 K are found to be 0.61, 0.48 and 0.45, respectively.

7.4.3 Discussion and conclusions

ClAlPc thin films were prepared by thermal evaporation. The dark conductivity of the films was measured in vacuum after deposition and shown to exhibit intrinsic semiconductor behaviour with a band-gap energy of 1.3 eV. The thin films were found to be very stable and no hysteresis was observed after temperature cycling between room temperature and 200°C. The films also exhibited reversible increases in dark conductivity on exposure to either O_2 or NO_2, although the response to the latter was much stronger and it was shown that it is possible to readily detect 500 ppm NO_2 in the presence of 20% O_2. These gases are adsorbed on the ClAlPc film surface where they induce surface acceptor states such that the ClAlPc can be considered to be doped by O_2 and NO_2 which are

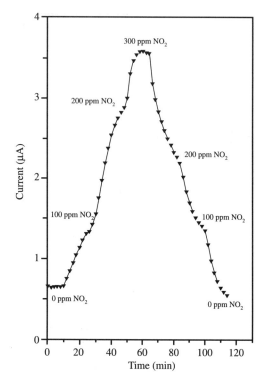

Figure 7.11(a)   Variation of current vs time for a ClAlPc thin film gas sensor on exposure to NO_2 in increasing/decreasing concentration steps (0, 100, 200 and 300 ppm) at 390 K

both oxidising gases. The corresponding acceptor level ionisation energies were determined to be 0.15 eV and 0.18 eV obtained for NO_2 and O_2, respectively, using the analysis of Roberts and Schmidlin. These values are in good agreement with previous findings by other workers for NO_2 adsorption on other phthalocyanines. We also examined the response of ClAlPc to NO_2 at different temperatures. The temperature of maximum conductivity (T_{max}) was determined to be 90°C for ClAlPc in response to NO_2 exposure. The reversibility of the response is not complete at this temperature. At high temperatures the adsorption and desorption processes are accelerated and the reversibility of the thin film sensor becomes more reproducible above a minimum temperature of operation (T_{min}) which was near 140°C in this case. The sensitivity (s) of our ClAlPc films to NO_2 was calculated and found to be 0.61 at 390 K; it becomes lower at higher temperatures.

T_{max} could be used as a design parameter for a thin film gas sensor since a maximum response is expected at this temperature. This temperature is different for different gases. ClAlPc thin films possess some

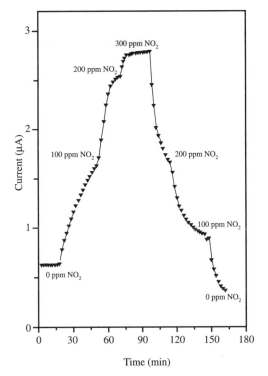

Figure 7.11(b)   Variation of current with time for a ClAlPc thin film gas sensor on exposure to NO_2 in increasing/decreasing concentration steps (0, 100, 200 and 300 ppm) at 420 K

desirable properties as a sensor material for toxic gases (i.e. thermal and chemical stability, sensitivity, reversibility and also simplicity/low cost).

The parameters T_{max}, T_{min} and s must therefore be carefully considered and the operating temperature chosen accordingly in order to obtain a gas sensor with optimum characteristics. The concentration dependence of the film sensitivity was found to be non-linear for each of the temperatures used here and no simple analytical expression for the concentration behaviour was derived. The long term stability and reproducibility of our ClAlPc is promising. Thin film sensors which were operated at elevated temperatures for several days showed no signs of material evaporation and, furthermore, many of our samples were successfully recycled by annealing them at 180°C in a vacuum for > 10 h in order to degas them. After this procedure they exhibited the same sensitivity to NO_2 exposure as before. However, unfortunately, the response speed of the thin films used in this study was rather slow and needs to be improved further before these ClAlPc thin films could be used as gas sensors in real industrial applications for NO_2 detection.

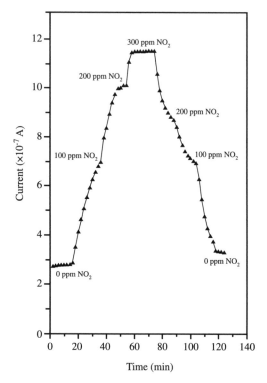

Figure 7.11(c)   Variation of current vs time for a ClAlPc thin film gas sensor on exposure to NO₂ in increasing/decreasing concentration steps (0, 100, 200 and 300 ppm) at 455 K

In this chapter we have given considerable experimental details of the practicality and usefulness of ClAlPc as an amperometric gas sensor, and have emphasised the care that must be taken in a real situation to ensure reliable results especially when detecting critical concentrations (sometimes in parts per billion, ppb) of toxic or highly inflammable gases. To conclude, we should like to refer to two strands of current work:

(i) As stated earlier, the phthalocyanines are promising for a whole range of gas sensing systems and one important property is being carefully investigated. The flat structure of the molecule makes it possible for the central atom, hydrogen or a metal, to be substituted or changed during the sample preparation and, furthermore, the radicals at the periphery of the molecule may also be substituted by a range of different radicals. This makes it possible to prepare a wide range of different, in some cases very complex phthalocyanines which are capable of a high degree of specificity. In other words, such a sensor can be made to respond to a particular gas but not to others mixed with it. One such material has been able to sense the presence of a few parts per million of toluene di-iso-cyanate, a noxious gas which accompanies some polymer processing systems. For health and safety purposes, such a sensor is invaluable.

(ii)   In Chapter 3 reference is made to various optical systems including the Mach–Zehnder interferometer. Such devices can be produced in miniature form and the surface sensitive to a light beam can be coated with a specially prepared phthalocyanine film. Gas absorbed by the phthalocyanine can be detected by an interrogating light beam, as the light absorption is modified by the absorbed gas. Such a system is exemplified in a recent paper by Hassan *et al.* [33] and has the advantage that no electrical connection to the detector is necessary. The probe is the light beam.

7.5 References

1   ELEY, D. D.: 'Phthalocyanines as semiconductors', *Nature*, 1948, **162**, p. 819

2   ABASS, A. K., KRIER, A., and COLLINS, R. A. :'The influence of iodine on the electrical properties of lead phthalocyanine (PbPc) thin film gas sensors', *Physica Status Solidi*, 1994, **12**, p. 345

3   RURHUA, W. U., and JONES, J. A.: 'The characteristics of gas-sensitive multilayer devices based on lead phthalocyanine', *Sensors and Actuators B*, 1990, **12**, p. 33

4   GOULD, R. D.: 'Dc electrical measurements on evaporated thin-films of copper phthalocyanine', *Thin Solid Films*, 1985, **125**, p. 63

5   SALEH, A. M., GOULD, R. A., and HASSAN, A. K.: 'Dependence of ac electrical parameters on frequency and temperature in zinc phthalocyanine thin-films', *Physica Status Solidi*, 1993, **139**, p. 379

6   AZIM-ARAGHI, M. E., CAMPBELL, D., KRIER, A., and COLLINS, R.A.: 'Electrical conduction mechanisms in thermally evaporated lead phthalocyanine thin films', *Semiconductor Science and Technology*, 1996, **11**, p. 39

7   WILSON, A., and COLLINS, R. A.: 'Electrical characteristics of planar phthalocyanine thin-film gas sensors', *Sensors and Actuators*, 1987, **12**, p. 389

8   HEILAND, G., and KOHL, D.: 'Problems and possibilities of oxidic and organic semiconductor gas sensors', *Sensors and Actuators*, 1985, **8**, p. 227

9   DOGO, S., GERMAIN, J. P., MALEYSSON, C., and PAULY, A.: 'Interaction of NO_2 with copper phthalocyanine thin-films. 1. Characterization of the copper phthalocyanine films', *Thin Solid Films*, 1992, **219**, p. 251

10 PASSARD, M., MALEYSSON, C., PAULY, A., DOGO, S., GERMAIN, J. P., and BLACE, J. P.: 'Gas sensitivity of phthalocyanine thin-films', *Sensors and Actuators B*, 1994, **18/19**, p. 489

11 BIRRELL, S. T., PEDLEY, D. G., PROSSER, S. J., WEBB, B. C., HIJIKIGAWA, M., and WADA, T.: 'European conference on sensors and their applications', European conference on *Sensors and their applications*, UMIST, Manchester, 1983

12 JANATA, J., and HUBER, R. J.: 'Principles of chemical sensors', *Ion-Selective Electrode Reviews*, 1979, **1**, p. 31

13 BOUSSE, L., DE ROOIJ, N. F., and BERGVELD, P.: 'Operation of chemically sensitive field-effect sensors as a function of the insulator–electrolyte interface', *IEEE Transactions on Electron Devices*, 1983, **ED-30**, (10), p. 1263

14 DANNETUM, H. M., PETERSON, L. G., SODERBERG, D., and LUNDSTORM, I.: 'A hydrogen sensitive pd-MOS structure working over a wide pressure range', *Applications of Surface Science*, 1984, **17**, p. 259

15 TAGUCHI, N., US Patent 3, Vol. 631, 1970, p. 436

16 WATSON, J.: 'Gas monitoring instrument utilising fibre optic, piezoelectric and gas-sensitive polymer techniques', *Sensors and Actuators*, 1984, **5**, p. 29

17 LALAUZE, R., and PIJOLAT, C.: 'Gas detection for automotive pollution control', *Sensors and Actuators*, 1984, **5**, p. 63

18 ROBERTSON, J. M.: 'An X-ray study of the phthalocyanines PII, Quantitative structure determination of the metal-free compound', *Journal of the Chemical Society*, 1936, **1**, p. 1195

19 BELGHACHI, A., and COLLINS, R. A.: 'Humidity response of phthalocyanine gas sensors', *Journal of Physics D, Applied Physics*, 1988, **21**, p. 1646

20 NAPIER, A., and COLLINS, R. A.: 'Phase-behavior of halogenated metal phthalocyanines', *Physica Status Solidi A*, 1994, **144**, p. 111

21 CAMPBELL, D. S.: 'Use of thin films in physical investigations' (Academic Press, New York, 1966)

22 CHOPRA, K. L.: 'Thin film phenomena' (McGraw-Hill, New York, 1969)

23 MAISSEL, L. I., and GLANGY, R.: 'Handbook of thin film technology' (McGraw-Hill, New York, 1970)

24 COLLINS, R. A., and MOHAMMED, K. A.: 'Gas sensitivity of some metal phthalocyanines', *Journal of Physics D, Applied Physics*, 1988, **21**, p. 154

25 UKEI, K.: 'Lead phthalocyanine', *Acta Crystallographica, Section B*, 1973, **29**, p. 2290

26 IYACHICA, Y., YAKUSHI, K., IKEMOTO, I., and KURODA, H.: 'Structure of lead phthalocyanine (triclinic form)', *Acta Crystallographica*, 1982, **38**, p. 766

27 GASTONGUAY, L., and VELLEUS, G.: 'Improvement of the photoelectrochemical activity of chloroaluminum phthalocyanine films by anion uptake and structural modifications', *Journal of the Electrochemical Society*, 1992, **139**, p. 337

28 BELGHACHI, A., and COLLINS, R. A.: 'The effects of humidity on phthalocyanine NO_2 and NH_3 sensors', *Journal of Physics D, Applied Physics*, 1990, **23**, p. 223

29 MEIER, H.: 'Organic semiconductors chemistry' (Weinheim, 1974)

30 BURR, P. M., JEFFERY, P. D., BENJAMIN, J. D., and UREN, M. J.: 'A gas-sensitive field-effect transistor utilizing a thin-film of lead phthalocyanine as the gate material', *Thin Solid Films*, 1987, **151**, p. L111

31 ROBERTS, G. G., and SCHMIDLIN, F. W.: 'Study of localised levels in semi-insulators by combined measurements of thermally activated ohmic and space-charge-limited conduction', *Physical Review*, 1969, **180**, p. 785

32 BOTT, B., and JONES, T. A.: 'A highly sensitive NO_2 sensor based on electrical-conductivity changes in phthalocyanine films', *Sensors and Actuators*, 1984, **5**, p. 43

33 HASSAN, A. K., RAY, A. K., TRAVIS, J. R., GHASSEMLOY, Z., COOK, M. J., ABASS, A., and COLLINS, R. A.: 'The effect of NO_2 on optical absorption in Langmuir–Blodgett films of octa-substituted amphiphilic copper phthalocyanine molecules', *Sensors and Actuators*, 1998, **B49**, p. 235

Index